Automotive Software Architectures

Miroslaw Staron

Automotive Software Architectures

An Introduction

 Springer

Miroslaw Staron
Department of Computer Science
    and Engineering
University of Gothenburg
Gothenburg, Sweden

ISBN 978-3-319-86441-9   ISBN 978-3-319-58610-6   (eBook)
DOI 10.1007/978-3-319-58610-6

Printed on acid-free paper

This Springer imprint is published by Springer Nature
The registered company is Springer International Publishing AG
The registered company address is: Gewerbestrasse 11, 6330 Cham, Switzerland

To my family—Sylwia, Alexander, Viktoria and Cornelia

Foreword

"Without exception, our aim must be to improve the current status; and instead of being satisfied with what has been achieved, we must always strive to do our job even better." This is the direction which the famous automotive entrepreneur Robert Bosch gave already at the dawn of the automotive age. It is still valid today, and we are indeed never satisfied with where we are in the automotive sector.

Software and IT are the major drivers of modern cars—both literally and from a marketing perspective. Modern vehicles have more than 50 electronic control units (ECUs), with premium cars having more than 100 such embedded computer systems. Some functions, such as engine control or dynamics, are hard real-time functions, with reaction times going down to a few milliseconds. Practically all other functions, such as infotainment, demand at least soft real-time behaviors.

Today automotive software is spearheading IT innovation. Software engineering for automotive systems today encompasses modern embedded and cloud technologies, distributed computing, real-time systems, mixed safety and security systems, and, last but not least, the connection of all these elements to long-term sustainable business models. The everyday relevance of automotive software for today's software engineers is high, and it is the focus of this book to bring this message to practitioners. Its main goal is to underline the convergence of embedded software with highly complex distributed IT systems.

Each automotive area has its own requirements for computational speed, reliability, security, safety, flexibility, and extensibility. Automotive electronic systems map functions such as braking, powertrain, or lighting controls to individual software systems and physical hardware. The resulting complexity has reached a limit that demands an architectural restart. At the same time, innovative functions such as connectivity with external infrastructures and vehicle-to-vehicle communication demand IT backbone and cloud solutions with service-oriented architectures (SOA).

Software and IT in vehicles and their environments are evolving at a fast pace. Multimodal mobility will connect previously separated domains like cars and public transportation. Mobility-oriented services such as car sharing creates completely new eco-systems and business models far away from the classic "buy your own car" approach. Autonomous driving demands highly interactive services with multi-

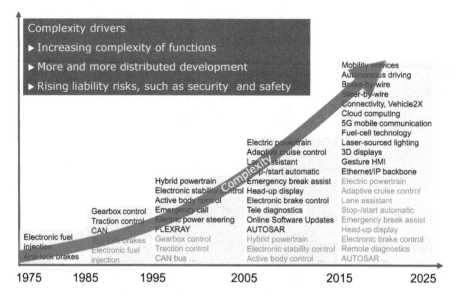

Fig. 1 Software and IT advance automotive innovations and complexity

sensor fusion, far away from currently deployed functionally isolated control units. Connectivity and infotainment have transformed the car into a distributed IT system with cloud access, over-the-air functional upgrades, and high-band-width access to map services, media content, other vehicles and surrounding infrastructure. Energy efficiency evolves the classic powertrain towards high voltage hybrid and electric engines.

With so many software-driven innovations in vehicles and their environments, complexity is growing fast—and in some cases beyond what can be controlled as recent security attacks have showed. Figure 1 indicates the rapid growth of software-driven innovations along with a forecast for the near future.

To master this fast growing complexity, automotive software needs a clear architecture. Architecture evolution today is the major focus across companies, and thus the book arrives just at the right time. Architecture impacts are manifold, such as systems modeling, testing and simulation with models in the loop; the combination of several quality requirements such as safety; service-oriented advanced operating systems with secure communication platforms such as adaptive AUTOSAR (Automotive Open System Architecture); multisensor fusion and picture recognition for ADAS (Advanced Driver Assistance Systems) and autonomous driving; distributed end-to-end security for flexible remote software updates directly into the cars' firmware; connectivity of cloud technologies and IT backbones with billions of cars and their on-board devices for infotainment, online apps, remote diagnosis and emergency call processing.

This book comprehensively introduces to automotive software architecture. Authored by renowned expert Miroslaw Staron it provides a guided tour through the methodology and usage of automotive software architecture. Starting with a brief introduction to software architecture paradigms it quickly moves to current application domains, such as AUTOSAR. Architecture analysis with methods such as ATAM of SEI provide hands-on guidance specifically at the current fast paradigm change from classic networking controllers to the three-tier model of future automotive IT.

With this book Miroslaw Staron and his co-authors target both engineers and decision-makers in the automotive electronics and IT domain. They guide engineers, developers and managers along the convergence of the two worlds of IT and embedded systems. Education however has only in rare cases dedicated programs for engineering this convergence of IT and embedded systems. Business models will evolve towards flexible service-oriented architectures and eco-systems. Reference points based on industry standards such as three-tier cloud architectures, adaptive AUTOSAR, and Ethernet connectivity facilitate reuse across companies and industries. The classic functional split is replaced by a more service-oriented architecture and delivery model. Development in the future will be a continuous process which will fully decouple the rather stable hardware of the car from its functionality driven by software upgrades. Hierarchic modeling of business processes, functionality and architecture from a systems perspective allow early simulation while ensuring robustness and security. Agile service delivery models combining DevOps, micro-services and cloud solutions will allow functional changes far beyond the traditional V approach.

The techniques presented in this book are not supposed to be the ultimate truth, but provide direction in a fast evolving field. It will help you as well as your organization to grow your maturity. Our society and each of us depend on seamless mobility, and so we need to also trust these underlying systems of infrastructure and vehicles. Let's evolve the necessary technology, methods, and competences in a good direction to stay in control of automotive software and avoid the many pitfalls of classic IT systems. For this reason I wish this book all the best and good success.

As with all architecture independent of application domain, we should not forget the wisdom of another great leader, Winston Churchill, who once said: "However beautiful the strategy, you should occasionally look at the results."

Stuttgart, Germany   Christof Ebert
February 2017

Preface

Even since I've learned how to drive, I've been an enthusiast of cars and driving. The ability to "go places" and be in charge of the machine that can take us to these places has always been something that I've loved. When I entered the field of computer science and software engineering, the software was present in cars in very few places—basically only to control the ignition of the engine. At least that was the case in the cars I owned at the time. However, I saw a large potential for using computers in cars.

It is the ability to use more software in cars that triggered my interest in automotive software architectures. In 2015 my publisher contacted me and proposed writing a book about topics I like. I managed to convince my colleagues—Darko Durisic from Volvo Car Group, Per Johannessen from AB Volvo and Wilhelm Meding from Ericsson—to help in writing some of the chapters.

In 2017 we managed to finish the book and we hope that it will provide a solid ground for our readers in designing automotive software. We hope that by writing this book we can contribute to a more exciting, yet at the same time safer, cars. We have enjoyed writing the book and we hope that you, our reader, will enjoy reading our book.

The purpose of the book is to introduce the concept of software architecture as one of the cornerstones of software in modern cars. The book is a result of my work in the area of software engineering, with particular focus on safety systems and software measurement. Throughout my research, I've worked with multiple companies in the automotive and telecom domains and I have noticed that over time these domains became increasingly similar. The processes and tools for developing software in modern cars became very similar to those used in the development of telecommunication systems. The same is very true about software architectures—initially very different, today they are increasingly similar in terms of architectural styles, programming paradigms and architectural patterns.

The book starts with a historical overview of the evolution of software in modern cars and the description of the main challenges which drive the evolution. Chapter 2 describes the main architectural styles of automotive software and their use in car's software. In Chap. 3, the reader can find a description of software development processes used to develop software on the car manufacturer's side. Chapter 4 introduces AUTOSAR—an important standard in automotive software. Chapter 5 goes beyond simple architecture and describes the process of detailed design of automotive software with the use of Simulink, which helps us understand how the detailed design links to the high-level design. Chapter 6 presents a method for assessing the quality of the architecture—ATAM (Architecture Trade-off Analysis Method)—and provides an example assessment. Chapter 7 presents an alternative way of assessing the architecture, namely by using quantitative measures and indicators. In Chap. 8 we dive deeper into one of the specific properties discussed in Chap. 6—safety—and can read about the important standard in that area—ISO/IEC 26262. Finally, Chap. 9 presents a set of future trends that seem to emerge today that have the potential to shape automotive software engineering in the coming years.

Gothenburg, Sweden   Miroslaw Staron
January 2017

Acknowledgements

First and foremost, I would like to thank the co-authors of some of the chapters in this book—Darko Durisic, Per Johannessen and Wilhelm Meding. I have had the privilege of working with them for a number of years and I'm deeply thankful for their insights into the car and telecom industries.

I am greatly indebted to my family—Sylwia, Alexander, Viktoria and Cornelia—who support me in taking on challenges and see to it that I am successful. They are the most fantastic family one could imagine.

I would also like to thank my publisher—Ralf Gerstner from Springer—who has proposed the idea of the book and helped me throughout the process. Without his encouragement and practical pointers this book would have never happened.

Many thanks to dSpace GmbH, for permitting me to use images of their equipment as part of the book. I also thank Jan Söderberg from Systemite for providing me with the figures and explanations on how the SystemWeaver tool keeps the different construction artifacts together.

I am grateful to my colleagues from Volvo Car Group who have taught me about practicalities of the automotive industry. I have met many persons from the fantastic team of Volvo Cars and had many great discussions about how cars are designed today, but in particular I am indebted to Kent Niesel, Martin Nilsson, Niklas Baumann, Anders Svensson, Hans Alminger, Ilker Dogan, Lars Rosqvist, Sajed Miremari, Mikael Sjöstrand and Peter Dahlslund. I would also like to thank Mark Hirche and Malin Folke for their comments on the draft of the book.

I would also like to thank my colleagues from the research community for their help and support in both writing this book and in my research activities leading to this book. In particular I would like to thank Imed Hammouda for his feedback and comments on the ATAM evaluation chapter.

And finally I would like to thank the Swedish Innovation Agency Vinnova, the Swedish Strategic Research Foundation SSF and the Software Center for providing me with research funding that allowed me to pursue my research interests in the area of this book.

Contents

Chapter 1
Introduction

Abstract Modern cars have evolved from mechanical devices into distributed cyber-physical systems which rely on software to function correctly. Starting from the 1970s the amount of electronics and software used has gradually increased from as little as one computer (Electronic Control Unit, ECU) to as much as 150 ECUs in 2015. The trend in the architecture, however, changes as companies look for ways to decrease the number of central computing nodes and connect them with the increased number of I/O nodes. In this chapter we provide an overview of the book and the conventions used in it and introduce the examples which we will use throughout. We describe the history of the automotive software anchoring the events in the evolution of the market of the electronics and software in modern cars. Towards the end of the chapter we also describe which directions can be pursued to deepen the knowledge of automotive software.

1.1 Software and Modern Cars

The introduction of software to cars opened up plenty of opportunities—from the optimization of cars' performance and to exciting infotainment features. Modern cars are full of electronics and the consumers are looking for car platforms which fully resemble software products. A good example of this kind of car is Tesla, which is known for innovations driven by software. The manufacturer is known for constantly pushing new versions of software to customers, providing them with new, exciting features almost every day.

The software intensive systems in modern cars provide plenty of new opportunities, but they also require more careful design, implementation, verification and validation before they can be released to users. And although the practices of software engineering include methods and tools able to fulfill the needs for safety and reliability of the automotive software, they must be applied in an automotive-specific manner to address these needs.

We could see the clear development of the automotive industry into a field less dominated by mechanical engineering but with a growing component of electronic and software engineering. We have seen the evolution of software from simple engine control algorithms of the 1970s to the advanced safety systems of the 2000s and the advanced connectivity of the 2010s. We can observe that the trends of using

© Springer International Publishing AG 2017
M. Staron, *Automotive Software Architectures*,
DOI 10.1007/978-3-319-58610-6_1

the software is not going to decrease, but will increase and the amount of software used will continue to increase.

With the growing amount and importance of software in modern cars we can observe the increased need for professional software engineering. Rigorous processes of software engineering lead to higher quality software with complexity not higher than necessary and assuring that the software does not contribute to fatalities in the traffic conditions.

One of the practices of software engineering is the high-level design of software systems, also referred to as *software architecture*. The architecture of the software provides the designers with the possibility to prescribe how the software functions are distributed to software components and how the components are to interact with each other. Software architecting is usually done at the early stages of software development and serves as the basis for the allocation of software modules to components and the distribution (called *systemization*) of the functions to software components.

1.2   History of Software in the Automotive Industry

Although today it is a given that there is a lot of software in our cars, it was not like that at the beginning of the automotive industry. The first cars did not contain any electronics, which only entered the automotive market during the 1970s with the introduction of electronic fuel injection as a response to the demand for fuel efficiency [CC11].

In the 1970s the software in the cars was usually embedded deeply in the electronics in functions related to single domains—e.g., electronic fuel injection in the powertrain, electronic ignition in the electrical system or central locking. Since the use of electronics was scarce in that decade, the notion of functional safety did not relate to software and it was relatively easy to embedded mechanisms for controlling the safety of the functions. The architectures of the software were usually monoliths which were not communicating with other parts of the software.

It was the 1980s that brought in such innovations as the central computers which could display basic telemetry of the vehicles—such as current fuel consumption, average fuel consumption and distance travelled. The ability to display the information to the drivers opened up new possibilities. On the embedded software front, software algorithms controlled new functions such as anti-lock brakes (ABS) and even electronic gearboxes.

The 1990s introduced even more consumer-visible electronics. The most notable innovation was in the infotainment domain and was the navigation system—or as it is commonly called, the GPS. Visualizing the information online required integration of important electronic components such as powertrain control computer, the dedicated GPS receiver and the infotainment display. The same decade introduced also more electronics and software in safety-critical areas such as ACC

Fig. 1.1 Late 1990s JECS LH-Jetronic ECU for engine control

(Adaptive Cruise Control) which controlled the speed of a vehicle based on the speed of the vehicles in front. The introduction of this kind of functionality raised the important questions of liability for accidents caused by malfunctioning of software. The automotive software architecture used in the 1990s was more distributed and software became often recognized as important factor in innovation in the car industry. An example computer system is presented in Fig. 1.1.[1]

This kind of development continued into the 2000s, when software started to dominate innovation in the car industry. It was also during the 2000s that the notion of advanced driver support systems was coined. The "advanced" referred to functions which integrated multiple computers in the car and made more "difficult" decisions for the driver. One of the most notable systems in this area was the City Safety system introduced by Volvo in its XC60 model [Ern13]. The system could stop the car from of speed under 50 kph when an obstacle appeared in front of it and the driver had no time to react. It was these kinds of systems that required more control over the complex interactions and prioritizations and therefore led to more advanced software architectures. The AUTOSAR standard was introduced to provide the possibility to communize solutions (where possible) and make it easy to change hardware platform with limited effort to adopt the software, and to enable easier sharing of the components between manufacturers and introduce a common "operating system" for the car's computers [Dur15, DSTH14].

---

[1] Author: RB30DE via Wikipedia https://en.wikipedia.org/wiki/JECS, under the Creative Commons License: http://creativecommons.org/licenses/by-sa/3.0/.

Fig. 1.2 2014 Audi TT infotainment unit

Finally, the 2010s introduced a completely new way of designing the electronics in cars [SLO10, RSB$^+$13]. Departing from the distributed network of computers in a single car, this decade introduced the concepts of wireless cars, car-2-car communication, car-2-infrastructure communication and autonomous driving concepts. Many new actors appeared on the market where the car was no longer a final product, but a platform where new functions could be deployed even post-production. Examples of such cars are Tesla cars or Google's self-driving vehicle [Mar10]. It was also this decade that required more advanced control over the execution of software coming from different vendors for the possibility of adding new functionality to cars without the need for physically modifying the cars. An example of a focus area—infotainment—is presented in Fig. 1.2.[2]

Another example is the infotainment unit of Volvo XC90 as presented in Fig. 1.3.

In today's cars the size of the software grows to over 100 million lines of code according to Viswanathan [Vis15].

---

[2]Author: Audi, available at https://en.wikipedia.org/wiki/JECS, under the Creative Commons License: http://creativecommons.org/licenses/by-sa/2.0/.

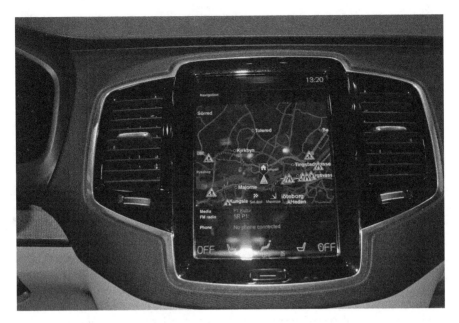

Fig. 1.3 2016 Volvo XC90 infotainment unit

1.3 Trends Shaping Automotive Software Development

In 2007, Pretschner et al. [PBKS07] outlined the major trends in software development in automotive systems. This work has been a trendsetter since then and has foreshadowed the large increase in the amount of automotive software—in 2007 measured in megabytes and in 2016 measured in gigabytes. The five trends of automotive software systems presented by Pretscher et al. are:

- Heterogeneity of software—the software in modern cars realizes different functions in different domains. These domains range from highly safety-critical (e.g. active safety) to user experience-centered (e.g. infotainment). This means that the ways of specifying, designing, implementing and verifying the software vary among domains.
- Distribution of labor—the development of the software systems is often distributed between automotive OEMs (Original Equipment Manufacturers, like Volvo, BMW, and Audi) and suppliers. Suppliers are also often given an option to define their own way of working as long as they comply with the requirements of and contracts with the OEMs.
- Distribution of software—the automotive software system comprises a number of ECUs, and each of the computers has its own software which needs to cooperate with other ECUs to fulfill its functions. This entails more difficulty in coordination of the software and introduces more complexity.

- Variants and configurations—the globalized and highly competitive automotive market requires customizations of the same car based on the requirements of the country and the user. This means that the software in modern cars needs to be able to work in different countries without the need for recertification and, therefore the software needs to handle variants in multiple ways—both in the source code and also at runtime.
- Unit-based cost models—the competitive market means that the unit price of the car cannot be too high compared to the competition and therefore it is often the case that automotive OEMs optimize the hardware and software in such a way that unit costs remains low while the development costs can be higher.

A lot has happened since 2007 and the major trends in the automotive market today can be complemented with such trends as[3]:

- Connectivity and cooperation [BWKC16]—the ability to use internet functions through mobile networks enabled cars to connect to each other and/or to use information from the infrastructure to make decisions. Research projects in the area of intelligent transport systems explore such ideas as planning of the speed of a bus to minimize the need for braking for "red" when approaching intersections. The modern cars are expected to be able to connect to smartphones via bluetooth and to use internet features such as web browsers or music services.
- Autonomous functions [LKM13]—the ability of the car to brake, steer and autonomously take over from drivers entails a large amount of complexity in safety-critical systems, but is seen as "the next big thing" in the automotive sector. This also means that the verification and validation methods for software in cars will become even more stringent and even more advanced.

Autonomous driving scenarios are challenging because of the need to have an accurate and exact model of the physical surroundings of the car. This demand for the accuracy requires more sophisticated measurement equipment and therefore more data to process, more decision points, and in turn more complex algorithms. One piece of such measurement equipment which is used in autonomous driving is LIDAR, shown in Fig. 1.4.[4]

Figure 1.4 shows a LIDAR mounted on the roof of an autonomous car. The device provides a 360-degree view of the surroundings and allows the car's software to find objects in the vicinity of the car. A LIDAR is often a complement to a RADAR, which is usually placed in the front of the vehicle. Figure 1.5 shows the picture of the radar ECU of a Volvo FH16 truck.

The production cars, however, do not have LIDARs yet, but take advantage of cameras placed in covered places. In Fig. 1.6 we can see the front camera of a Volvo XC90.

---

[3]Based on author's own observations.

[4]Author: Steve Jurvetson; available at flickr.com, under the Creative Commons License: http://creativecommons.org/licenses/by/2.0/.

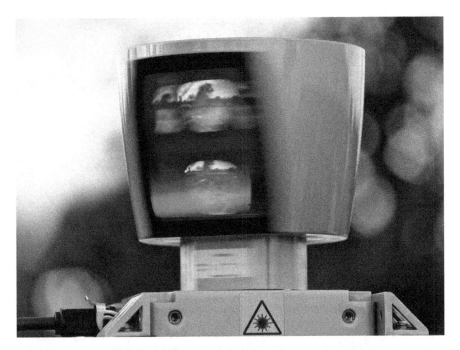

Fig. 1.4 Velodyne High-Def LIDAR

Fig. 1.5 Radar ECU in Volvo FH16 truck

It is interesting to observe the automotive software market today, and therefore we believe that this book will be of use to anyone who is interested in starting to get into automotive software engineering.

Fig. 1.6 Front camera in Volvo XC90

1.4 Organization of Automotive Software Systems

Over the years each car manufacturer (often referred to as an OEM, Original
Equipment Manufacturer) developed its own way of organizing software systems
with the diversity in pair of the diversity of car brands today. However, many of the
car manufacturers design the software in a similar way—they use the V development
model and a similar organization of the electrical (and software) systems into
domains and subsystems. We can depict it in the model presented in Fig. 1.7.

In this view we can see that the electrical system is organized into domains,
such as infotainment and powertrain. Each of these domains has a specific set of
properties—some are safety-critical and some not, some are very user oriented and
some are realtime and embedded. Each of these domains, however, is organized into
subsystems which group a specific functionality (some OEMs call these subsystems
simply "systems") such as active safety, and advanced driver support and similar.
These systems group a number of logical elements and realize the functionality,
which is often grouped into functions. The functions are often called end-to-end
functions, as they realize user functionality such as Adaptive Cruise Control, Line
Departure Warning and Navigation from A to B.

The functions are realized by subsystems of the electrical system and they are
orthogonal to the organization of subsystems, components and modules. Therefore
we often see the concept of "functional architecture (view)"—describing the
dependencies among functions.

Each subsystem contains a number of components which include smaller parts
of software elements that realize parts of the functionality (e.g. such a part

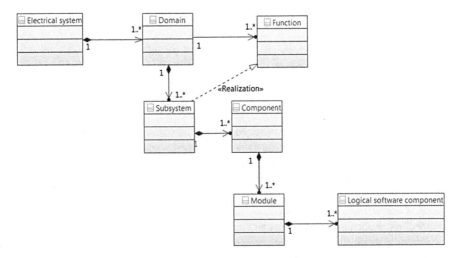

Fig. 1.7 Conceptual view of the organization of the software system

could be a message broker for an infotainment system). These components are organized into software modules, which are often source code files with a set of classes, methods and programming language functions. The groupings of these programming language functions or software classes are referred to as logical software components.

The term software architecture can be used in almost all levels of this hierarchy (except for the lowest one). We can talk about the EE architecture (Electrical System architecture) which describes the organization of software and hardware for the entire car. We can talk about an ECU architecture which describes the logical organization of software subsystems, components and modules in the ECU. Depending on the size and role of the ECU we could have modules, components or subsystems in the ECU [DNSH13].

The methods and techniques presented in this book can be applied at any of these levels.

1.5   Architecting as a Discipline

Software architecture is a kind of artifact in software development, but architecting is a full-fledged discipline with its own activities and tasks. It is quite often the case that software architects are perceived as more experienced than senior designers and are given a larger mandate to make decisions than software designers. In order to prevent confusion, let us briefly discuss the role of software architects in contrast to the designers and project managers. These two roles can be perceived as overlapping to some extent and therefore this comparison gets interesting.

1.5.1   Architecting vs. Project Management

Being a software architect means being in a role of a kind of technology leadership. The architects are the persons who lay the ground for the development of the entire system—in terms of general architectural styles, but also in terms of principles which guide the development of the system. Those principles form the boundaries within which the designers can make their choices. It is the role of the architect to ensure that these principles are followed during the entire lifecycle of the software system.

In some sense, setting the frames for the system design is a technical correspondent to setting the frames for the cost and scope of the project that develops the system. However, it is the responsibility of the project manager to set and monitor this project scope, schedule and cost. Therefore we contrast architecting as a technical correspondent to project management in Table 1.1.

Since the discipline of architecting is practices by technical experts, it is technical principles that are applied—how to create objects, send messages, deploy components onto ECUs. This means that the technologies and their characteristics are in focus. For example, the architects need to balance different quality characteristics with each other—performance vs. safety, maintainability vs. portability and others. Therefore the architects also focus on the quality and functionality—addressing such challenges as "how to enable video feeds over the Flexray network without adding new cables". Finally the architects focus on the functionality and make sure that the electrical system of the car can realize the functionality given the constraints (e.g. weight of the cables, number of ECUs). All of these aspects make software architecting seem as technical product management.

In contrast to the technical management, we have project management, where the project leaders apply organizational theories to determine whether to work Agile or waterfall, or how to negotiate contracts, or how to measure the progress of the project. When applying the managerial and organizational theories the project leaders focus on the scope of the project—addressing the questions of whether a given functionality can be developed given the budget constraints of the project. The focus of the project leaders is on resources, on balancing cost and resources with the schedule of the project. All of these aspects can be seen as management of the project rather than management of the product.

Table 1.1 Architecting vs. project management

Architecting	Project management
Done by technical experts	Done by management experts
Technology in focus	Scope in focus
Focus on quality	Focus on cost
Focus on requirements	Focus on work products
Focus on solution	Focus on resources
Maximize functionality	Minimize cost

Table 1.2 Architecting vs. designing

Architecting	Designing
Making rules and decisions	Following rules and decisions
High level structures	Low-level structures
Holistic understanding	Specialistic understanding
Systems thinking	Software thinking
Documentation-oriented	Code and executable/detailed model-oriented
Modelling and analysis	Execution and testing

Both technical and project management need to work with one another as they develop the one and the same product! Humphrey [Hum96] in his book "Managing Technical People: Innovation, Teamwork and the Technical process" provides a number of useful guidelines on how to combine these two.

1.5.2   Architecting vs. Design

Similarly to contrasting the discipline of architecting to the discipline of project management, we can also contrast architecting to designing. We could observe from the previous contrast that technical product management is about setting principles for the work. The discipline of designing is all about following these principles in order to arrive at final software product. We present some of the differences in Table 1.2.

Software architecting, being the technical management of the system, sets the boundaries for the design in terms of principles, rules and decisions about how to design the system. An example of such a decision is the choice of the communication protocol between the ECUs and the number of ECUs in the system. It's also about which standards to follow and why. Architecting, as we will see in this book, is a discipline operating at a high abstraction level—considering components (e.g. groups of software classes) and execution nodes. This requires a holistic understanding of the system—both the software and the underlying hardware used to execute the software or provide the software with data. This kind of a "systems thinking" makes the architects the core part of any software team because they understand the background of "why" things happen rather than just do things.[5]

The discipline of architecting is also very documentation-oriented—as the decisions, rules and principles need to be communicated, they also need to be explained and documented to lead to consistency and enforcement of rules. This happens often as a process of analysis and modelling of the system.

---

[5]Sinek in his book "Starting with Why: How Great Leaders Inspire Everyone to Action" [Sin11] presents a set of examples of how this works in practice.

In contrast, the discipline of designing is focused on realizing the principles, decisions and rules of the architecture in software code or an executable model. The high-level structure discussed in the architecture is now developed using lower-level structures—components using classes and blocks, ECUs using execution processes. This requires specialized knowledge and competence in the particular domain in question (e.g. the infotainment or powertrain). The design is focused on the software entities and their interaction with the underlying hardware, where the hardware is often given (or at least the specification of the hardware is given during the design of the software). This means that designing is focused on the code and executable/detailed models rather than on abstract analysis and modelling. It is also therefore the design that is the first activity where we discuss testing and execution, whereas in the architecture we talk about assessments and evaluations (a topic which we will return to in Chap. 6).

Similarly to the collaboration between the architects and the project managers, the architects need to collaborate closely with the designers in order to develop and deliver a software system which fulfills all the requirements and quality constraints.

1.6 Content of This Book

This book addresses one of the most fundamental aspects of engineering of software systems—software architectures. The architecture is a high-level design of a software system which enables the architects to distribute the functionality of the software system to multiple interacting components. The components are usually grouped into subsystems and domains which address a set of functional and non-functional requirements of the software system.

In this book we explore the concept of software architecture for modern cars which is intended for both novice and advanced software designers. This book is intended for two groups of audience—professionals working with automotive software who need to understand concepts related to automotive architectures, and students of software engineering or related programs who need to understand the specifics of automotive software to be able to construct cars or their components.

The idea to support the professionals came from the author's observations that the automotive industry requires an individual software engineer to be able to understand a variety of disciplines. Individuals working with the construction of car software or hardware need to understand their counterparts in order to be able to design safe, reliable and long-term solutions for the car industry. Software engineers need to understand how their software is to be integrated with other software from other vendors in order to be able to develop user functions, e.g. collision avoidance by braking.

The idea to support the students came from the observation that many of the graduates from software engineering programs require further education in order to understand such advanced concepts as software and systems safety, working with suppliers and distribution of software. During the author's years of working with students it became evident that it is difficult to provide education in software engineering in general and also focus on specific aspects such as automotive software. This book addresses this challenge and is aimed at being both a reference book and a potential course book for software engineering programs.

This book is structured into independent chapters which can be read separately, although we recommend reading them in sequence. Reading the chapters in sequence allows us to follow the motivating example throughout the book and to gradually build up knowledge about automotive software architectures.

1.6.1 Chapter 2: Software Architectures

In this chapter we present the basics of software architecture in general as a recap for readers who are not familiar with architecting as a discipline, and towards the end of the chapter we describe the specificity of automotive software architectures.

In the beginning of the chapter we review the definitions of software architectures, define the types of view used in automotive software design and relate them to the architectural views in software engineering in general—the 4+1 architecture view model.

We gradually progress in the chapter to introduce elements important for automotive architectures, e.g., ECUs (Electronic Control Units), logical and physical components, functional architectures, and topologies for automotive architectures (physical and logical). We delve into the peculiarities of automotive software—embedded systems with large focus on safety and dependability.

1.6.2 Chapter 3: Automotive Software Development

In this chapter we describe and elaborate on software development processes in the automotive industry. We introduce the V-model for the entire vehicle development and we continue to introduce modern agile software development methods for describing the ways of working of software development teams. We also provide an overview of a tool which is used to keep the design data consistent—SystemWeaver by SystemIte.

In this chapter we discuss the specifics of automotive software development such as variant management, different integration stages, testing strategies and the methods used for these. We review methods used in practice and explain how they should be used.

1.6.3 Chapter 4: AUTOSAR Reference Model

In this chapter we continue on the topic of standardization and we discuss the current standardization efforts. We describe and discuss the AUTOSAR standard, which gets the most attention today in Europe and worldwide.

In the AUTOSAR standard we describe the main building blocks like software components and communication buses. We also describe the evolution of the standard from the perspective of the main concepts and their influence on the car industry.

Towards the end of the chapter we present the AUTOSAR reference architecture as described in the standard and discuss its evolution.

1.6.4 Chapter 5: Detailed Design of Automotive Software

In this chapter we continue to delve into the technical aspects of automotive software architectures and we describe ways of working when designing software within particular software components. We present the methods for modelling the functions using Simulink modelling and we show how these methods are used in the automotive industry.

Towards the end of the chapter we introduce the need for quality assessment of software architectures and the challenges related to assessment of the sub-characteristics of quality (the so-called "-ilities").

1.6.5 Chapter 6: Evaluation of Automotive Software
Architectures

In this chapter we introduce methods for assessing the quality of software architectures and we discuss ATAM. We discuss the non-functional properties of automotive software and we review the methods used to assess such properties as dependability, robustness and reliability. We follow the ISO/IEC 25000 series of standards when discussing these properties.

In this chapter we also address the challenges related to the integration of hardware and software and the impact of this integration. We review the differences with stand-alone desktop applications and discuss examples of these differences.

Towards the end of the chapter we discuss the need to measure these properties and introduce the need for software measurement.

1.6.6 Chapter 7: Metrics for Software Design and Architectures

In this chapter we describe the most commonly used metrics in software engineering in general and in automotive software engineering, e.g. lines of code, model size, complexity, and architectural stability or coupling [SHFMHNH13]. In particular we present these metrics and their interpretation (what should be done, and why, based on the values of metrics). We discuss the use of metrics based on the international standard ISO/IEC 15939.

1.6.7 Chapter 8: Functional Safety of Automotive Software

In this chapter we elaborate on one of the most important issues related to software in modern cars—functional safety. We explore the safety-related concepts described in the international standard ISO/IEC 26262 and we describe how this standard is used in modern software development processes.

We explore such elements as verification and validation techniques mentioned in the standard and link them to the ASIL levels and efficiency of their applications.

In the chapter we describe how the standard is to be applied on the examples of the simple function.

1.6.8 Chapter 9: Current Trends in Automotive Software Development

We conclude the book with the outlook on the current trends in automotive software development and we introduce the emerging, disruptive technologies on the market that have the potential to change the automotive industry to become more software-oriented than it traditionally has been.

1.6.9 Motivating Examples in the Book

In this book we illustrate the concepts introduced in each chapter with a set of examples. Each chapter has its own examples which are dedicated to extrapolating the concepts described, and therefore:

- Chapter 2 contains a set of examples from different domains, e.g. infotainment, powertrain and active safety.

- Chapter 3 includes examples of requirements from AUTOSAR and requirements for opening the car from the chassi domain.
- Chapter 4 contains examples of the AUTOSAR models and their realization for communication between two ECUs.
- Chapter 5 includes examples of digitalization of an analog signal and the designing of the heating of a car's chassi from the Chassi domain.
- Chapter 6 contains examples of the parking assistance camera from the active safety domain.
- Chapter 7 contains examples of a real software (obfuscated) published as open source.
- Chapter 8 includes the example of a simple microcontroller demonstrating the different ASIL levels and architectural choices used to achieve these levels.

These examples do not constitute an entire software system of a car, as these systems are huge. As a reference, BMW in its talks at conferences showed the size of the electrical system to be about 200 ECUs, which includes all variants of its electrical system (meaning that there is no car with all 200 ECUs.[6])

1.7   Knowledge Prerequisites

In order to understand the book one needs to understand how programming works. We do not require any specific programming skills, but it is good to know the basics of programming in C/C++ or Java/C#. It is also good to have the basic knowledge of the UML notation, especially the class diagrams.

We introduce topics from the automotive domain and we require no prior understanding of the domain nor any knowledge of software architecture.

For each chapter we provide pointers where the interested reader can find more information or where the necessary prerequisites can be obtained.

1.8   Where to Go Next

After reading this book you will be able to understand how to architect a software system for a modern car. You will also be prepared to understand the design principles guiding the development of software in modern cars and be able to understand the non-functional principles behind the design.

The next natural step is to follow your interest in the design of software systems. We recommend focusing on the principles of continuous integration and deployment, virtual verification and validation as well as advanced functional safety.

---

[6]Presentation from BMW at Elektronik i Fordon, Gothenburg, May 2016.

References

BWKC16.   Robert Bertini, Haizhong Wang, Tony Knudson, and Kevin Carstens. Preparing a roadmap for connected vehicle/cooperative systems deployment scenarios: Case study of the state of oregon, usa. *Transportation Research Procedia*, 15:447–458, 2016.

CC11.   Andrew YH Chong and Chee Seong Chua. *Driving Asia: As Automotive Electronic Transforms a Region*. Infineon Technologies Asia Pacific Pte Limited, 2011.

DNSH13.   Darko Durisic, Martin Nilsson, Miroslaw Staron, and Jörgen Hansson. Measuring the impact of changes to the complexity and coupling properties of automotive software systems. *Journal of Systems and Software*, 86(5):1275–1293, 2013.

DSTH14.   D. Durisic, M. Staron, M. Tichy, and J. Hansson. Evolution of Long-Term Industrial Meta-Models - A Case Study of AUTOSAR. In *Euromicro Conference on Software Engineering and Advanced Applications*, pages 141–148, 2014.

Dur15.   D. Durisic. *Measuring the Evolution of Automotive Software Models and Meta-Models to Support Faster Adoption of New Architectural Features*. Gothenburg University, 2015.

Ern13.   Tomas Ernberg. Volvo's vision 2020–'no death, no serious injury in a volvo car'. *Auto Tech Review*, 2(5):12–13, 2013.

Hum96.   Watts S Humphrey. *Managing technical people: innovation, teamwork, and the software process*. Addison-Wesley Longman Publishing Co., Inc., 1996.

LKM13.   Jerome M Lutin, Alain L Kornhauser, and Eva Lerner-Lam MASCE. The revolutionary development of self-driving vehicles and implications for the transportation engineering profession. *Institute of Transportation Engineers. ITE Journal*, 83(7):28, 2013.

Mar10.   John Markoff. Google cars drive themselves, in traffic. *The New York Times*, 10(A1):9, 2010.

PBKS07.   Alexander Pretschner, Manfred Broy, Ingolf H Kruger, and Thomas Stauner. Software engineering for automotive systems: A roadmap. In *2007 Future of Software Engineering*, pages 55–71. IEEE Computer Society, 2007.

RSB$^+$13. Rakesh Rana, Miroslaw Staron, Christian Berger, Jörgen Hansson, Martin Nilsson, and Fredrik Törner. Increasing efficiency of iso 26262 verification and validation by combining fault injection and mutation testing with model based development. In *ICSOFT*, pages 251–257, 2013.

SHFMHNH13. Staron, M., Hansson, J., Feldt, R., Meding, W., Henriksson, A., Nilsson, S. and Höglund, C., 2013, October. Measuring and visualizing code stability–a case study at three companies. In *The International Conference on Software Process and Product Measurement*, (pp. 191–200). IEEE.

Sin11.   Simon Sinek. *Start with why: How great leaders inspire everyone to take action*. Penguin UK, 2011.

SLO10.   Margaret V String, Nancy G Leveson, and Brandon D Owens. Safety-driven design for software-intensive aerospace and automotive systems. *Proceedings of the IEEE*, 98(4):515–525, 2010.

Vis15.   Balaji Viswanathan. Driving into the future of automotive technology at genivi annual members meeting. *OpenSource Delivers*, online, 2015.

Chapter 2
Software Architectures: Views and Documentation

Abstract Software architecture is the foundation for automotive software design. Being a high-level design view of the system it combines multiple views on the software system, and provides the project teams with the possibility to communicate and make technical decisions about the organization of the functionality of the entire software system. It allows also us to understand and to predict the performance of the system before it is even designed. In this chapter we introduce the definitions related to software architectures which we will use in the reminder of the book. We discuss the views used during the process of architectural design and discuss their practical implications.

2.1 Introduction

As the amount of software in modern cars grows we observe the need to use more advanced software engineering methods and tools to handle the complexity, size and criticality of the software [Sta16, Für10]. We increase the level of automation and increase the speed of delivery of software components. We also constantly evolve software systems and their design in order to be able to keep up with the speed of the changes in requirements in automotive software projects.

Software architecture is one of the cornerstones of successful products in general, and in particular in the automotive industry. In general, the larger the system, the more difficult it is to obtain a good quality overview of its functions, subsystems, components and modules—simply because of the limitations of our human perception. In automotive software design we have more specific challenge, related to the safety of the software embedded in the car and the distribution of the software—both distribution in terms of the physical distribution of the computing nodes and distribution of the development among the car manufacturers and their suppliers.

In this chapter we discuss the concept of software architecture and explain it with the examples of building architectures. Once we know more about what constitutes software architecture, we go into the details of different views of software architecture and how they come together. We then move on to describing the most common architectural styles and explain where they can be seen in automotive software. Finally we present the ways of describing architectures—the

© Springer International Publishing AG 2017
M. Staron, *Automotive Software Architectures*,
DOI 10.1007/978-3-319-58610-6_2

architecture modelling languages. We end the chapter with references to further readings for readers interested in more details.

2.2 Common View on Architecture in General and in the Automotive Industry in Particular

The concept of architecture is well rooted in our society and its natural association is to the styles of buildings. When thinking about architecture we often recall large cathedrals, the gothic and modern styles of churches, or other large structures. One of the examples of such a cathedral is the "Sagrada Familia" cathedral in Barcelona with its very characteristic style.

However, let us discuss the concept of the architecture with a considerable smaller example—let us take the example of a pyramid. Figure 2.1[1] presents a picture of the pyramids in Gizah.

The form of the pyramid is naturally based on a triangle. The fact that it is based on a triangle is one of the architectural choices. Another choice is the type of the triangle (e.g. using the golden number 1.619 as the ratio between the slant height to half the base length). The decision is naturally based on mathematics and illustrated

Fig. 2.1 All Gizah pyramids: a picture represents an external view of the product

---

[1] Author: Ricardo Liberato, available at Wikipedia, under the Creative Commons License: https://creativecommons.org/licenses/by-sa/2.0/.

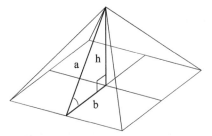

Fig. 2.2 Internal view of the architecture of a pyramid

Fig. 2.3 Volvo XC 90, another example of the external view of the product

using one of the views of the pyramid—call it an early design blueprint as presented in Fig. 2.2.

Figure 2.2 shows the first design principles later on used to detail the design of the pyramid. Instead of delving deeper into the pyramid construction, let us now consider the notion of architecture and software architecture in the automotive industry.

One obvious view of the architecture of the car is the external view of the product, as with the view of the pyramid (Fig. 2.3).[2]

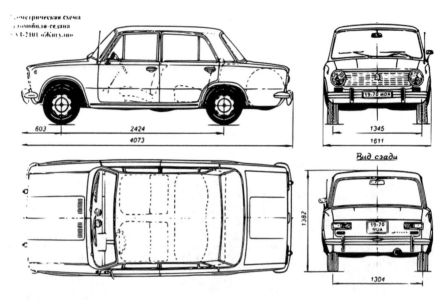

Fig. 2.4 A blueprint of the design principles of a car

We can observe the general architectural characteristics of a car—the placement of the lights, the shape of the lights, the shape of the front grill, the length of the car, etc. This view has to be complemented with a view of the internal design of the car. An example of such a blueprint is presented in Fig. 2.4.[3]

This blueprint shows the dimensions of the car, hiding other kinds of details. Another blueprint can be a view of the electrical system of a car, its electronics and its software.

2.3 Definitions

Software architecting starts with the very first requirement and ends with the last defect fix in the product, although its most intensive period is in the early design stage where the architects decide upon the high-level principles of the system design. These high-level principles are documented in the form of a software architecture document with several views included. We could therefore define the software architecture as the high-level design, but this definition would not be just. The definition which we use in this book is:

---

[3]Figure source: pixbay.com.

Software architecture refers to the high-level structures of a software system, the discipline of creating such structures, and the documentation of these structures. These structures are needed to reason about the software system

The definition is not the only one, but it reflects the right scope of the architecture. The definition comes from Wikipedia (https://en.wikipedia.org/wiki/Software_architecture).

2.4 High-Level Structures

The definition presented in this chapter ("Software architecture refers to the high-level structures of a software system...") talks about "high-level structures" as a means to generalize a number of different entities used in the architectural design. In this chapter we go into details about these structures, which are:

1. Software components/Blocks—pieces of software packaged into subsystems and components based on their logical structure. Examples of such components could be UML/C++ classes, C code modules, and XML configuration files.
2. Hardware components/Electronic Control Units—elements of design of the computer system (or platform) on which the software is executed. Examples of such elements include ECUs, communication buses, sensors and actuators.
3. Functions—elements of the logical design of the software described in terms of functionality, which is then distributed over the software components/blocks. Examples of such elements are software functions, properties and requirements.

All of these elements together form the electrical system of the car and its software system. Even though the hardware components do not "belong" to the software world, it is the often the job of the architect to make sure that they are visible and linked to the software components. This linking is important from the process perspective—it must be know which supplier should design the software for the hardware. We talk more about the concept of the supplier and the process in Chap. 3.

In the list of high-level structures, when introducing functions, we indicated the interrelation between these entities—"functions distributed over the software components". This interrelation leads us to an important principle of architecting—the use of views. An architectural view is *a representation of one or more structural aspects of an architecture that illustrates how the architecture addresses one or more concerns held by one or more of its stakeholders* [RW12].

One could see the process of architecting as a prescriptive design, the process continuous as the design evolves. Certain aspects of design decisions influence the architecture and are impossible to know a priori—increased processing power required to fulfill late function requirements or safety-criticality of the designed system. If not managed correctly the architecture has a tendency to evolve into a descriptive documentation that needs to be kept consistent with the software itself [EHPL15, SGSP16].

2.5 Architectural Principles

The second part of the definition of the software architecture ("...the discipline of creating such structures...") refers to the decisions which the software architects make in order to set the scene for the development. The software architects create the principles by defining such things as what components should be included in the system, which functionality each component should have (but not how it should be implemented—this is the role of the design discipline, which we describe in Chap. 5) and how the components should communicate with each other.

Let us consider the coupling as an example of setting the principles. We can consider an example of a communication between the component representing the controller of the windshield wipers and the component representing the hardware interface to the small engine controlling the actual windshield wiper arm. We could have a coupling in one way, as presented in Fig. 2.5.

In the figure we can see that the line (association) between the blocks is directed from WindshieldWiper to WndEngHW. This means that the communication can only happen in one way—the controller can send signals to the hardware interface. This seems logical, but it raises challenges when the controller wants to know the status of the hardware interface without pulling the interface—it is not possible as the hardware interface cannot communicate with the controller. If an architect sets this principle then this has the consequences on the later design, such as the need for extra signals on the communication bus (pulling the hardware for the status).

However, the software architect might make another decision—to allow communication both ways, which is shown in Fig. 2.6.

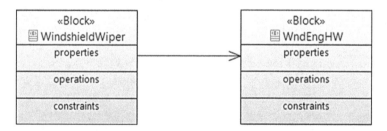

Fig. 2.5 An example principle—unidirectional coupling between two blocks

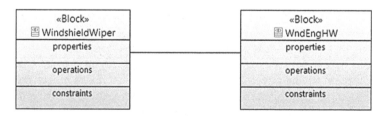

Fig. 2.6 An example principle—bidirectional coupling between two blocks

The second architectural alternative allows the communication in both ways, which solves the challenges related to pulling the hardware interface component for the status. However, it also brings in another challenge—tight coupling between the controller and the hardware interface. This tight coupling means that when one of these two component changes, the other should be changed (or at least reviewed) as the two are dependent on one another.

In the reminder of this chapter we discuss several of such principles when discussing architectural styles.

2.6   Architecture in the Development Process

In order to put the process of architecting in context and describe the current architectural views in automotive software architectures, let us first discuss the V-model as shown in Fig. 2.7. The V-model represents a high-level view of a software development process for a car from the perspective of OEMs. In the most common scenario, where there is no OEM in-house development, component design and verification is usually entirely done by the suppliers (i.e., OEMs send empty software compositions to the suppliers, who populate them with the actual software components).

The first level is the functional development level, where we encounter two types of the architectural views—the functional view and the logical system view. Now, let us look into the different architectural views, their purpose and the principles of using them. When discussing the views we also discuss the elements of these views.

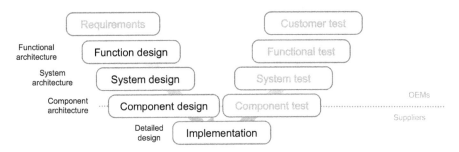

Fig. 2.7 V-model with focus on architectural views and evolution

2.7   Architectural Views

As we show in the process when starting with the development from scratch, the requirements of or ideas for functions in the car come first—the product management has the ideas about what kind of functionality the car should have. Therefore we start with this type of the view first and gradually move on to more detailed views on the design of the system.

2.7.1   Functional View

The functional view, often abbreviated to *functional architecture*, is the view where the focus is on the functions of the vehicle and their dependencies on one another [VF13]. An example of such a view is shown in Fig. 2.8.

As we can see from the example, there are three elements in this diagram—the functions (plotted as rounded-edge rectangles), the domains (plotted as sharp-edged rectangles) and the dependency relations (plotted as dashed lines), as the functions can depend on each other and they can easily be grouped into "domains" such as Powertrain and Active Safety. The usual domains are:

1. Powertrain—grouping the elements related to the powertrain of the car—engine, engine ECU, gearbox and exhaust.
2. Active Safety—grouping the elements related to safety of the car—ADAS (Advanced Driver Assistance Systems), ABS (Anti-lock Braking System) and similar.

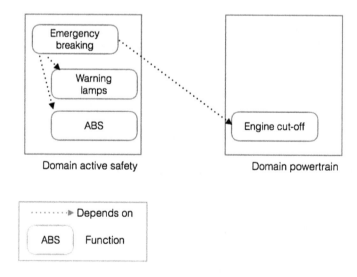

Fig. 2.8 Example of a functional architecture—or a functional view

3. Chassi and body—grouping the elements related to the interior of the car—seats, windows and other (which also contain electronics and software actuators/sensors).
4. Electronic systems—grouping the elements related to the functioning of the car's electronic system—main ECU, communication buses and related.

In modern cars the number of functions can reach more than 1000 and is constantly growing. The largest growth in the number of functions is related to new types of functionality in the cars—autonomous driving and electrification. Examples of functions from the autonomous driving area are:

1. Adaptive Cruise Control—basic function to automatically keep a distance from the preceding vehicle while maintaining a constant maximum velocity.
2. Lane Keeping Assistance—basic function to warn the driver when the vehicle is crossing the parallel line on the road without the turn indicator.
3. Active Traffic Light Assistance—medium advanced function to warn the driver of a red light ahead.
4. Traffic Jam Chauffeur—medium/advanced function to autonomously drive during traffic jam conditions.
5. Highway Chauffeur/pilot—medium/advanced function to autonomously drive during high-speed driving.
6. Platooning—advanced function to align a number of vehicles to drive autonomously in a so-called platoon.
7. Overtaking Pilot—advanced function to autonomously drive during an overtake situation.

These advanced functions build on top of the more basic functionality of the car, such as the ABS (Anti-lock Braking System), warning lights and blinkers. The basic functions that are used by the above functions can be exemplified by:

1. Anti-lock Braking System (ABS)—preventing the car from locking the brakes on slippery roads
2. Engine cut-off—shutting down the engine in situations such as after crash
3. Distance warning—warning the driver about too little distance from the vehicle in front.

The functional view provides the architects with the possibility to cluster functions, and distribute them to the right department to develop and to reason about these kinds of functionality. An example of such reasoning is the use of methods such as the Pareto front [DST15].

2.7.1.1 How-To

The process of functional architecture design starts with the development of the list of functions of the vehicle and their dependencies, which can be documented in block diagrams, use case diagrams or SysML requirements diagrams [JT13, SSBH14].

Once the list and dependencies are found, we move to organizing the functions to the domains. In the normal case these domains are known and given. The organization of the functions is based on how they are dependent on each other with the principle that the number of dependencies that cross-cut the domains should be minimized. The result of this process is the development of the diagram as shown in Fig. 2.8.

2.7.2   *Physical System View*

Another view is the system view on the architecture, usually portrayed as a view of the entire electrical system at the top level with accompanying lower-level diagrams (e.g. class diagrams in UML). Such an overview level is presented in Fig. 2.9. In this view we could see the ECUs (rounded rectangles) of different sizes placed on two physical buses (lines). This view of the architecture provides the possibility to present the topology of the electrical system of the car and provides the architects with a way to reason about the placement of the computers on the communication buses.

In the early days of automotive software engineering (up until the late 1990s) this view was quite simple and static as there were only a few ECUs and a few communication buses. However, in the modern software design, this view gets increased importance as the number of ECUs grows and the ability to give an overview becomes more important. The number of communication buses also increases and therefore the topologies of the components in the physical architectures have evolved from the typical star topologies (as in Fig. 2.9) to more linked architectures with over 100 active nodes. The modern view on the topology

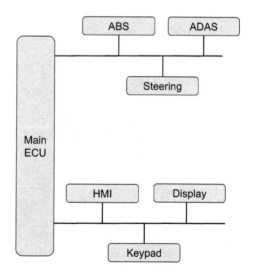

Fig. 2.9 Example of a system architecture—or a system view

also includes information about the processing power and the operating system (and its version) of each ECU.

2.7.2.1   How-To

Designing this view is usually straightforward as it is dictated by the physical architecture of the car, where the set of ECUs is often given. The most important ECUs are often predetermined from the previous projects—usually the main computer, the active safety node, the engine node, and similar. A list of the most common ECUs present in almost all modern cars is (https://en.wikipedia.org/wiki/Electronic_control_unit):

- Engine control unit (EnCU)
- Electric power steering control unit (PSCU)
- Human-machine interface (HMI)
- Powertrain control module (PCM)
- Telematic control unit (TCU)
- Transmission control unit (TCU)
- Brake control module (BCM; ABS or ESC)
- Battery management system

Depending on the car manufacturer, the other control modules can differ significantly. It is also the case that many of the additional control units are part of the electrical system, meaning that they are included only in certain car models or instances, depending on the customer order.

2.7.3   Logical View

The focus of the view is on the topology of the system. This view is often accompanied by the logical component architecture as presented in Fig. 2.10. The rationale behind the logical view of the system is to focus solely on the software of the car. In the logical view we show which classes, modules, and components are used in the car's software and how they are related to each other. The notation used for this model is often UML (Unified Modelling Language) and its subling SysML (Systems Modelling Language).

For the logical view, the architects often use a variety of diagrams (e.g. communication diagrams, class diagrams, component diagrams) to show various levels of abstraction of the software of the car—usually in its context. For the detailed design, these architectural models are complemented with low-level executable models such as Matlab/Simulink defining the behaviour of the software [Fri06].

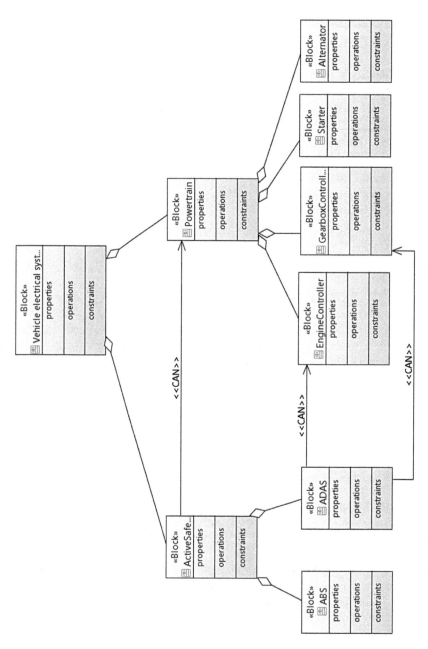

Fig. 2.10 Example of a logical view—a UML class diagram notation

2.7.3.1 How-To

The first step in describing the logical view of the software is to identify the components—these are modelled as UML classes. Once they are identified we should add the relationships between these components in the form of associations. It is important to keep the directionality of the associations correct as these will determine the communication between the components added during the detailed design.

The logical architecture should be refined and evolved during the entire project of the automotive software development.

2.7.4 Relation to the 4+1 View Model

The above-mentioned three views presently used in automotive software engineering evolved from the widely known principles of 4+1 view architecture model presented in 1995 by Kruchten [Kru95]. The 4+1 view model postulates describing software architectures from the following perspectives:

- logical view—describing the design model of the system, including entities such as components and connectors
- process view—describing the execution process view of the architecture, thus allowing us to reason about non-functional properties of the software under construction
- physical view—describing the hardware architecture of the system and the mapping of the software components on the hardware platform (deployment)
- development view—describing the organization of software modules within the software components
- scenario view—describing the interactions of the system with the external actors and internal interactions between components.

These views are perceived as connected with the scenario view overlapping the other four, as presented in Fig. 2.11, adapted from [Kru95].

The 4+1 view model has been used in the telecommunication domain, the aviation domain and almost all other domains. Its close relation to the early version of UML (1.1–1.4) and other software development notations of the 1990s contributed to its wide spread and success.

In the automotive domain, however, the use of UML is rather limited to class/object diagrams and therefore this view is not as common as in the telecommunication domain.

Fig. 2.11 4+1 view model of architecture

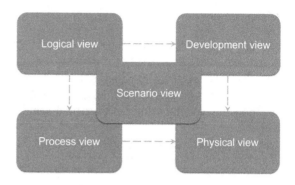

2.8 Architectural Styles

As the architecture describes the high-level design principles of the system, we can often observe how these design decisions shape the system. In this case we can talk about the so-called architectural styles. The architectural styles form principles of software design in the same way as building architecture shapes the style of a building (e.g. thick walls in gothic style).

In software design we distinguish between a number of styles in general, but in the automotive systems we can only see a number of those, as the automotive software has harder requirements on reliability and robustness than, for example, web servers. Therefore some of the styles are not applicable.

In this section, let us dive deeper into architectural styles and their examples.

2.8.1 Layered Architecture

This architectural style postulates that components of the system are placed in a hierarchy on top of each other and function calls (API usage) are made only from higher to lower levels, as shown in Fig. 2.12.

We can often see this type of layered architecture in the design of microcontrollers and in the upcoming AUTOSAR standard where the software components are given specific functions such as communication. An example of this kind of architecture is presented in Fig. 2.13.

A special variant of this kind of style is the two-tier style as presented by Steppe et al. [SBG+04], with one layer for the abstract components and the other one for the middleware details. One example of middleware can be found in Chap. 4 in the description of the AUTOSAR standard. Examples of the functionality implemented by the middleware are logging diagnostic events, handling communication on the buses, securing data and data encryption.

An example of such an architecture can be seen in the area of autonomous driving when dividing decisions into a number of layers, as shown in Fig. 2.14 extended from [BCLS16].

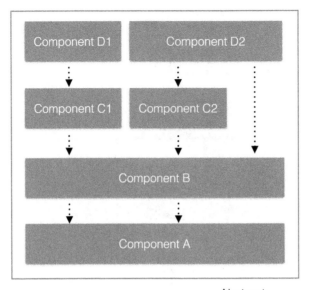

Fig. 2.12 Layered architectural style—*boxes* symbolize components and *lines* symbolize API usage/method calls

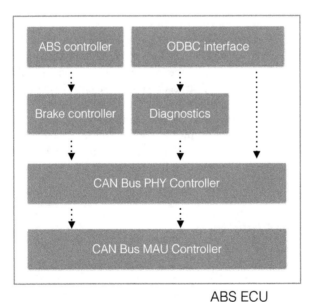

Fig. 2.13 An example of a layered architecture

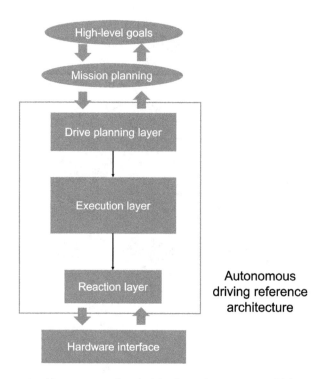

Fig. 2.14 Layered architecture example—decision layers in autonomous driving

In this figure we can see that the functionality is distributed in different layers and the higher layers are responsible for mission/route planning while the lower levels are responsible for steering the car. This kind of modular layered architecture allows the architects to distribute competence into the vertical domains. The blue arrows indicate that this architecture is abstract and that these layers can be connected either directly or indirectly (i.e. there may be other layers in-between).

We quickly realize the this kind of architectural style has limitations caused by the fact that the layers can communicate only in one way. The components within the same layer are often not supposed to communicate. Therefore, there is another style which is often used—component-based.

2.8.2 Component-Based

This architectural style is more flexible than the layered architecture style and postulates the principle that all components can be interchangeable and independent of each other. All communication should go through well-defined public interfaces

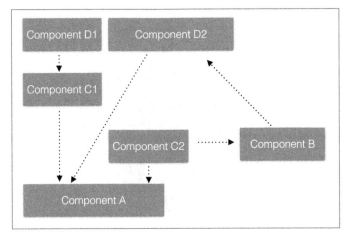

Abstract
representation of
the system

Fig. 2.15 Component-based architectural style

and each component should implement a simple interface, allowing for queries about which interfaces are implemented by the component. In the non-automotive domain this kind of architecture has been populated by Microsoft in its Windows OS through the usage of DLLs (Dynamic Linked Libraries) and the IUnknown interface.

An abstract view of this kind of style is presented in Fig. 2.15.

The component-based style is often used together with the design-by-contract principle, which postulates that the components should have contracts for their interfaces—what the API functions can and cannot do. This component-based style is often well suited when describing the functional architecture of the car's functionality.

In contemporary cars we can see this architectural style in the Infotainment domain, where the system is divided into the platform and the application layer (thus having layered architecture), and for the application layer all the apps which can be downloaded onto the system are designed according to component-based principles. These principles mean that each app can use another one as long as the apps have the right interface. For example a GPS app can use the app for music to play sound in the background without leaving the GPS. As long as the music app exposes the right interface, it makes no difference to the GPS app which music app is used.

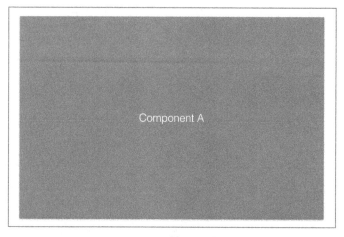

Abstract
representation of
the system

Fig. 2.16 Monolithic architectural style

2.8.3 Monolithic

This style is the opposite of that of component-based architecture as it postulates that the entire system is one large component and that all modules within the system can use each other. This style is often used in low-maturity systems as it leads to high coupling and high complexity of the system. An abstract representation in shown in Fig. 2.16.

The monolithic architecture is often used for implementing parts of the safety-critical system, where the communication between components needs to be done in real time with as little communication overhead as possible. Typical mechanisms in the monolithic architectures are the "safe" mechanisms of programming languages such as use of static variables, no memory management and no dynamic structures.

2.8.4 Microkernel

Starting in the late 1980s, software engineers started to use microkernel architecture when designing operating systems. Many of the modern operating systems are built in this architectural style. In short, this architectural style can be seen as a special case of the layered architecture with two layers:

- Kernel—a limited set of components with the higher execution privileges, such as task scheduler, memory manager, and basic interprocess communication

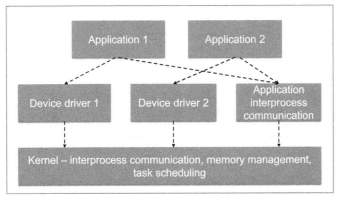

Fig. 2.17 Microkernel architectural style

manager. These components have the precedence over the application layer components.

- Application—components such as user application processes, device drivers, or file servers. These components can have different privilege levels, but always lower than that of the kernel processes.

The graphical overview of such an architectural style is show in Fig. 2.17.

In this architectural style it is quite common that applications (or components) communicate with each other over interprocess communications. This type of communication allows the operating system (or the platform) to maintain control over the communications.

In the automotive domain, the microkernel architecture is used in certain components which require high security. It is argued that the minimality of the kernel allows us to apply the principles of least privilege, and therefore remain in control of the security of the system at all times. It is also sometimes argued that hypervisors of the virtualized operating systems are build according to this principle. In the automotive domain the use of virtualization is currently in the research stage, but seems to be very promising as it would allow us to minimize the costs of hardware while at the same time retain the flexibility of the electrical system (imagine all cars had the same hardware and one could only use different virtual OSs and applications for each brand or type of car!).

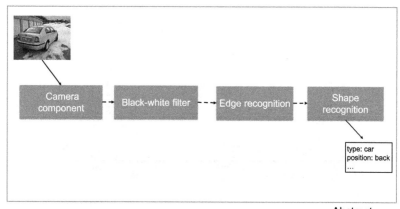

Fig. 2.18 Pipes and filters architectural style

2.8.5 Pipes and Filters

Pipes and filters is another well-known architectural style which fits well for systems that operate based on data processing (thus making its "comeback" as Big Data enters the automotive market). This architectural style postulates that the components are connected along the flow of the data processing, which is conceptually shown in Fig. 2.18.

In contemporary automotive software, this architectural style is visible in such areas as image recognition in active safety, where large quantities of video data need to be processed in multiple stages and each component has to be independent of the other (as shown in Fig. 2.18) [San96].

2.8.6 Client–Server

In client-server architectural style the principles of the design of such systems prescribe the decoupling between components with designated roles—servers which provide resources upon the request of the clients, as shown in Fig. 2.19. These requests can be done in either the pull or the push manner. Pulled requests mean that the responsibility for querying the server lies with the client, which means that the clients need to monitor changes in resources provided by the server. Pushed requests mean that the server notifies the relevant clients about changes in the resources (as in the event-driven architectural style and the published subscriber style).

In the automotive domain, this style is seen in specific forms like publisher-subscriber style or event-driven style. We can see the client-server style in such

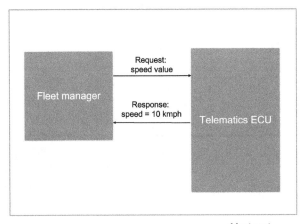

Fig. 2.19 Client-server architectural style

components as telemetry, where the telematics components provide the information to the external and internal servers [Nat01, VS02].

2.8.7 Publisher–Subscriber

The publisher–subscriber architectural style can be seen as a special case of the client–server style, although it is often perceived as a different style. This style postulates the principle of loose coupling between providers (publishers) of the information and users (subscribers) of the information. Subscribers subscribe to a central storage of information in order to get notifications about changes in the information. The publisher does not know the subscribers and the responsibility of the publisher is only to update the information. This is in clear contrast to the client–server architecture, where the server sends the information directly to a known client (known as it is the client that sends the request). The publisher–subscriber style is illustrated in Fig. 2.20.

In automotive software, this kind of architectural style is used when distributing information about changes in the status of the vehicle, e.g. the speed status or the tire pressure status [KM99, KB02]. The advantage of this style is the decoupling of information providers from information subscribers so that the information providers do not get overloaded as the number of subscribers increases. However, the disadvantage is the fact that the information providers do not have control of which components use the information and what information they possess at any given time (as the components do not have to receive updates synchronously).

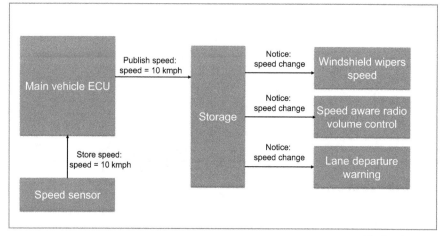

Abstract
representation of
the system

Fig. 2.20 Publisher–subscriber architectural style

2.8.8   Event-Driven

The event-driven architectural style has been popularized in software engineering together with graphical user interfaces and the use of buttons, text fields, labels and other graphical elements. This architectural style postulates that the components listen for (hook into) the events that are sent from the component to the operating system. The listener components react upon receipt of the event and process the data which has been sent together with the event (e.g. position of the mouse pointer on the screen when clicked). This is conceptually presented in Fig. 2.21.

The event driven architectural style is present in a number of parts of the automotive software system. Its natural placement with the user interface of the infotainment or the driver assist systems (e.g. voice control), which is also present in the aviation industry [Sar00] is obvious. Another use is diagnostics and storage of the error codes [SKM+10]. Using Simulink to design software systems and using stimuli and responses, or sensors and actuators, shows that event-driven style has been incorporated.

2.8.9   Middleware

The middleware architectural style postulates the existence of a common request broker which mediates the usage of resources between different components. The

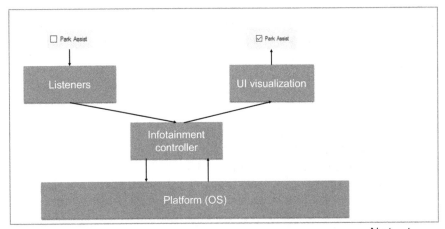

Abstract
representation of
the system

Fig. 2.21 Event-driven architectural style

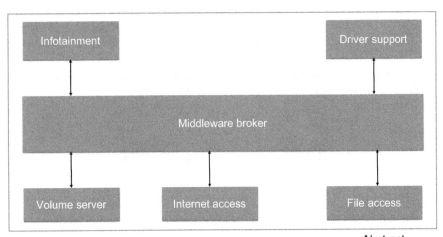

Abstract
representation of
the system

Fig. 2.22 Middleware architectural style

concept has been introduced into software engineering together with the initiative of CORBA (Common Object Request Broker Architecture) by Object Management Group [OPR96, Cor95]. Although the CORBA standard itself is not relevant for the automotive domain, its principles are present in the design of the AUTOSAR standard with its meta-model to describe the common elements of automotive software. The conceptual view of middleware style is shown in Fig. 2.22.

In automotive software, the middleware architecture is visible in the design of the AUTOSAR standard, which is discussed in detail later on in this book. The usage of middleware becomes increasingly important in automotive software's mechanisms of adaptation [ARC+07] and fault tolerance [JPR08, PKYH06].

2.8.10 Service-Oriented

Service-oriented architectural style postulates loose coupling between component using internet-based protocols. The architectural style puts emphasis on interfaces which can be accessed as web services and is often depicted as in Fig. 2.23.

Here the services can be added and changed on-demand during the runtime of the system.

In automotive software, this kind of architecture style is not widely used, but there are areas where the on-demand or ad hoc services are needed. One examples is vehicle platooning which has such an architecture [FA16], and is presented in Fig. 2.24.

Since vehicle platooning is done "spontaneously" during driving, the architecture needs to be flexible and needs to allow vehicles to link to and unlink from each other without the need to recompile or restart the system. The lack of available interfaces can lead to change in the vehicle operation mode, but not to disturbance in the software operation.

Now that we have introduced the most popular architectural styles, let us discuss the languages used to describe software architectures.

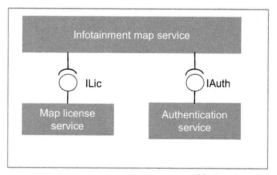

Fig. 2.23 Service-oriented architectural style

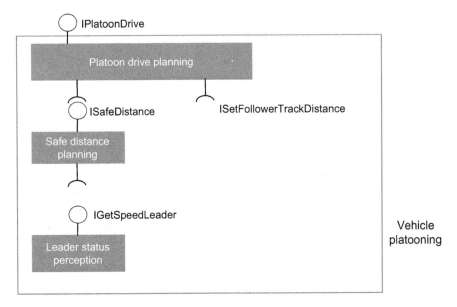

Fig. 2.24 An example of a service-oriented architecture—vehicle platooning

2.9 Describing the Architectures

In this book we have seen multiple ways of drawing architectural diagrams depending on the purpose of the diagram. We used the formal UML notation in Fig. 2.10 when describing the logical components of the software. In Fig. 2.8 we used boxes and lines, which are different from the boxes and lines used in Figs. 2.12, 2.13, 2.14, 2.15, 2.16, 2.17, 2.18, 2.19, 2.20, 2.21, and 2.22. It all has a purpose.

By using different notations we could see that there is no unified formalism describing a software architecture and that software architecture is a means of communication. It allows architects to describe the principles guiding the design of their system and discuss the implications of the principles on the components. Each of these notations could be called ADL—Architecture Description Language. In this section we introduce the most relevant ADLs which are available for software architects, with the focus on two formalisms—SySML (Systems Modelling Language, [HRM07, HP08]) and EAST-ADL [CCG+07, LSNT04].

2.9.1 SysML

SySML is a general-purpose language based on Unified Modelling Language (UML). It is built as an extension of a subset of UML to include more diagrams (e.g. Requirements Diagram) and reuse a number of UML symbols with the profile mechanism. The diagrams (views) included in SySML are:

- Block definition diagram—an extended class diagram from UML 2.0 using stereotyped classes to model blocks, activities, their attributes and similar. As the "block" is the main building block in SySML, it is reused quite often to represent both software and hardware blocks, components and modules.
- Internal block diagram—similar to the block definition diagram, but used to define the elements of a block itself
- Package diagram—the same as the package diagram from UML 2.0, used to group model elements into packages and namespaces
- Parametric diagram—diagram which is a special case of the internal block diagram and allows us to add constraints to the elements of the internal block diagram (e.g. logical constraints on the values of data processed).
- Requirement diagram—contains user requirements for the system and allows us to model and link them to the other model elements (e.g. blocks). It is one of the diagrams that adds a lot of expressiveness to SySML models, compared to the standard Use Case diagrams of UML.
- Activity diagram—describes the behaviour of the system as an activity flow.
- Sequence diagram—describes the interaction between block instances in a notation based on MSC (Message Sequence Charts) from the telecommunications domain.
- State machine diagram—describes the state machines of the system or its components.
- Use case diagram—describes the interaction of the system with its external actors (users and other systems).

An example of a requirement diagram is presented in Fig. 2.25 from [SSBH14].

The diagram presents two requirements related to each other (Maximum Acceleration and Engine Power) with the dependency between them. Blocks like the "6-Cylinder Engine" are linked to these requirements with the dependency "satisfy" to show where these requirements are implemented.

As we can quickly see from this example, the requirements diagram can be used very effectively to model the functional architecture of the electrical system of a car.

The block diagram was presented when discussing the logical view of the architecture (Fig. 2.10) and it can be further refined into a detailed diagram for a particular block, as shown in Fig. 2.26.

The diagram fulfills a similar purpose as the detailed design of the block, which is often done using the Simulink modelling language. In this book we look into the details of Simulink design in Chap. 6.

The behavioral diagrams of SySML are important for the detailed design of automotive systems, but they are out of the scope of this chapter as the architecture model is supposed to focus on the structure of the system and therefore kept on a high abstraction level.

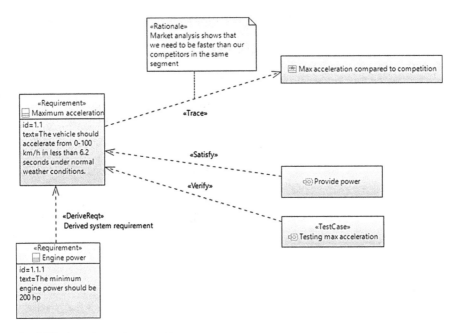

Fig. 2.25 Example requirements diagram

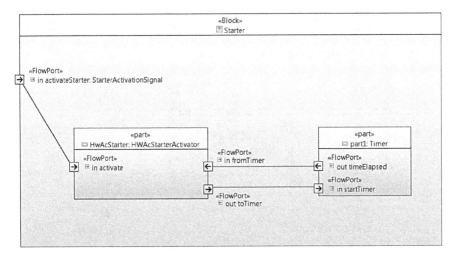

Fig. 2.26 Internal block diagram

2.9.2   EAST ADL

EAST ADL is another modelling language based on UML which is intended
to model automotive software architectures [CCG⁺07, LSNT04]. In contrast to
SySML, which was designed by an industrial consortium, EAST ADL is the result

of a number of European Union-financed projects which included both research and development components.

The principles of EAST ADL are similar to those of SySML in the sense that it also allows us to model automotive software architecture in different abstraction levels. The abstraction levels of EAST ADL are:

- Vehicle level—architectural model describing the vehicle functionality from an external perspective. It is the highest abstraction level in EAST ADL, which is then refined in the Analysis model.
- Analysis level—architectural model describing the functionality of the vehicle in an abstract model, including dependencies between the functions. It is an example of a functional architecture, as discussed in Sect. 2.7.1.
- Design level—architectural model describing the logical architecture of the software, including its mapping to hardware. It is similar to the logical view from Fig. 2.10.
- Implementation level—detailed design of the automotive software; here EAST ADL reuses the concepts from the AUTOSAR standard.

The vehicle level can be seen as a use case level of the specification where the functionality is designed from a high abstraction level and then gradually refined into the implementation.

Since EAST ADL is based on UML, the visual representation of models in EAST ADL is very similar to the models already presented in this chapter. However, there are some differences in the structure of the models and therefore the concepts used in SySML and EAST ADL may differ. Let us illustrate one of the differences with the requirements model in Fig. 2.27.

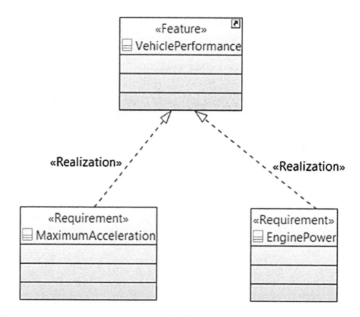

Fig. 2.27 Feature (requirements) diagram in EAST ADL

The important difference here is the link of the requirement—in EAST ADL the requirements can be linked to Features, a concept which does not exist in SySML.

In general, EAST ADL is a modelling notation more aligned with the characteristics of the automotive domain and makes it easier to structure models for a software engineer. However, EAST ADL is not as widely spread as SySML and therefore not as widely adopted in industry.

2.10 Next Steps

After the architecture is designed in the different diagrams, it should be transferred to the product development database and linked to all the other elements of the electrical system of the car. The product development database contains the design details of all software and hardware components, the relationships between them and the deployment of the logical software components onto the physical components of the electrical system.

2.11 Further Reading

The architectural views, styles and modelling languages, discussed in this section, are the most popular one used in the software industry today. However, there are also others, which we encourage the interested reader to explore.

Alternative modelling languages which are used in industry are the UML MARTE profile [OMG05, DTA$^+$08]. The MARTE profile has been designed to support modelling of real-time systems in all domains where they are applicable. Therefore there is a significant body of knowledge from using this profile, including executable variants of it [MAD09].

Readers interested in extending modelling languages can find more information in our previous work on language customization [SW06, SKT05, KS02, SKW04] and the way in which these extension can be taught [KS05].

An interesting review of future directions of architectures in general has been conducted by Kruchten et al. [KOS06]. Although the review was conducted over a decade ago, most of its results are valid today.

2.12 Summary

In this chapter we presented the concept of software architecture, its different viewpoints, and its architectural styles and introduced two notations used in automotive software engineering—SySML and EAST ADL.

An interesting aspect of automotive software architectures is that they usually mix a number of styles. The overall style of the architecture can be layered architecture within an ECU, but the architecture of each of the components in the ECU can be service-oriented, pipes and filters or layered. A concrete example is the AUTOSAR architecture. AUTOSAR provides a reference three layer architecture where the first "application" layer can implement service-oriented architecture, the second layer can implement a monolithic architecture (just RTE) and the third, "middleware", layer can implement component-based architecture.

The reasons for mixing these styles is that the software within a modern car has to fulfill many functions and each function has its own characteristics. For the telematics it is the connectivity which is important and therefore client-server style is the most appropriate. Now that we have discussed the basics of architectures, let us dive deeper into other activities in automotive software development, to understand why architecture is so important and what comes before and next.

References

ARC+07.   Richard Anthony, Achim Rettberg, Dejiu Chen, Isabell Jahnich, Gerrit de Boer, and Cecilia Ekelin.   Towards a dynamically reconfigurable automotive control system architecture. In *Embedded System Design: Topics, Techniques and Trends*, pages 71–84. Springer, 2007.

BCLS16.   Manel Brini, Paul Crubillé, Benjamin Lussier, and Walter Schön. Risk reduction of experimental autonomous vehicles: The safety-bag approach. In *CARS 2016 workshop, 4th International Workshop on Critical Automotive Applications: Robustness and Safety*, 2016.

CCG+07.   Philippe Cuenot, DeJiu Chen, Sebastien Gerard, Henrik Lonn, Mark-Oliver Reiser, David Servat, Carl-Johan Sjostedt, Ramin Tavakoli Kolagari, Martin Torngren, and Matthias Weber. Managing complexity of automotive electronics using the EAST-ADL. In *12th IEEE International Conference on Engineering Complex Computer Systems (ICECCS 2007)*, pages 353–358. IEEE, 2007.

Cor95.   OMG Corba. The common object request broker: Architecture and specification, 1995.

DST15.   Darko Durisic, Miroslaw Staron, and Matthias Tichy.   Identifying optimal sets of standardized architectural features – a method and its automotive application. In *2015 11th International ACM SIGSOFT Conference on Quality of Software Architectures (QoSA)*, pages 103–112. IEEE, 2015.

DTA+08.   Sébastien Demathieu, Frédéric Thomas, Charles André, Sébastien Gérard, and François Terrier. First experiments using the UML profile for MARTE. In *2008 11th IEEE International Symposium on Object and Component-Oriented Real-Time Distributed Computing (ISORC)*, pages 50–57. IEEE, 2008.

EHPL15.   Ulf Eliasson, Rogardt Heldal, Patrizio Pelliccione, and Jonn Lantz. Architecting in the automotive domain: Descriptive vs prescriptive architecture. In *Software Architecture (WICSA), 2015 12th Working IEEE/IFIP Conference on*, pages 115–118. IEEE, 2015.

FA16.   Patrik Feth and Rasmus Adler. Service-based modeling of cyber-physical automotive systems: A classification of services. In *CARS 2016 workshop, 4th International Workshop on Critical Automotive Applications: Robustness and Safety*, 2016.

Fri06.   Jon Friedman. MATLAB/Simulink for automotive systems design. In *Proceedings of the conference on Design, Automation and Test in Europe*, pages 87–88. European Design and Automation Association, 2006.

Für10.   Simon Fürst. Challenges in the design of automotive software. In *Proceedings of the Conference on Design, Automation and Test in Europe*, pages 256–258. European Design and Automation Association, 2010.

HP08.   Jon Holt and Simon Perry. *SysML for systems engineering*, volume 7. IET, 2008.

HRM07.   Edward Huang, Randeep Ramamurthy, and Leon F McGinnis. System and simulation modeling using SysML. In *Proceedings of the 39th conference on Winter simulation: 40 years! The best is yet to come*, pages 796–803. IEEE Press, 2007.

JPR08.   Isabell Jahnich, Ina Podolski, and Achim Rettberg. Towards a middleware approach for a self-configurable automotive embedded system. In *IFIP International Workshop on Software Technolgies for Embedded and Ubiquitous Systems*, pages 55–65. Springer, 2008.

JT13.   Marcin Jamro and Bartosz Trybus. An approach to SysML modeling of IEC 61131-3 control software. In *Methods and Models in Automation and Robotics (MMAR), 2013 18th International Conference on*, pages 217–222. IEEE, 2013.

KB02.   Jörg Kaiser and Cristiano Brudna. A publisher/subscriber architecture supporting interoperability of the can-bus and the internet. In *Factory Communication Systems, 2002. 4th IEEE International Workshop on*, pages 215–222. IEEE, 2002.

KM99.   Joerg Kaiser and Michael Mock. Implementing the real-time publisher/subscriber model on the controller area network (can). In *2nd IEEE International Symposium on Object-Oriented Real-Time Distributed Computing, 1999*, pages 172–181. IEEE, 1999.

KOS06.   Philippe Kruchten, Henk Obbink, and Judith Stafford. The past, present, and future for software architecture. *IEEE software*, 23(2):22–30, 2006.

Kru95.   Philippe B Kruchten. The 4 + 1 view model of architecture. *Software, IEEE*, 12(6):42–50, 1995.

KS02.   Ludwik Kuzniarz and Miroslaw Staron. On practical usage of stereotypes in UML-based software development. *the Proceedings of Forum on Design and Specification Languages, Marseille*, 2002.

KS05.   Ludwik Kuzniarz and Miroslaw Staron. Best practices for teaching uml based software development. In *International Conference on Model Driven Engineering Languages and Systems*, pages 320–332. Springer, 2005.

LSNT04.   Henrik Lönn, Tripti Saxena, Mikael Nolin, and Martin Törngren. Far east: Modeling an automotive software architecture using the east adl. In *ICSE 2004 workshop on Software Engineering for Automotive Systems (SEAS)*, pages 43–50. IET, 2004.

MAD09.   Frédéric Mallet, Charles André, and Julien Deantoni. Executing AADL models with UML/MARTE. In *Engineering of Complex Computer Systems, 2009 14th IEEE International Conference on*, pages 371–376. IEEE, 2009.

Nat01.   Martin Daniel Nathanson. System and method for providing mobile automotive telemetry, July 17 2001. US Patent 6,263,268.

OMG05.   UML OMG. Profile for modeling and analysis of real-time and embedded systems (marte), 2005.

OPR96.   Randy Otte, Paul Patrick, and Mark Roy. *Understanding CORBA: Common Object Request Broker Architecture*. Prentice Hall PTR, 1996.

PKYH06.   Jiyong Park, Saehwa Kim, Wooseok Yoo, and Seongsoo Hong. Designing real-time and fault-tolerant middleware for automotive software. In *2006 SICE-ICASE International Joint Conference*, pages 4409–4413. IEEE, 2006.

RW12.   Nick Rozanski and Eóin Woods. *Software systems architecture: Working with stakeholders using viewpoints and perspectives*. Addison-Wesley, 2012.

San96.   Keiji Saneyoshi. Drive assist system using stereo image recognition. In *Intelligent Vehicles Symposium, 1996., Proceedings of the 1996 IEEE*, pages 230–235. IEEE, 1996.

Sar00.   Nadine B Sarter. The need for multisensory interfaces in support of effective attention allocation in highly dynamic event-driven domains: the case of cockpit automation. *The International Journal of Aviation Psychology*, 10(3):231–245, 2000.

SBG⁺04.   Kevin Steppe, Greg Bylenok, David Garlan, Bradley Schmerl, Kanat Abirov, and
          Nataliya Shevchenko. Two-tiered architectural design for automotive control systems:
          An experience report. In *Proc. Automotive Software Workshop on Future Generation
          Software Archtiecture in the Automotive Domain*, 2004.

SGSP16.   Ali Shahrokni, Peter Gergely, Jan Söderberg, and Patrizio Pelliccione.   Organic
          evolution of development organizations – An experience report. Technical report, SAE
          Technical Paper, 2016.

SKM⁺10.   Chaitanya Sankavaram, Anuradha Kodali, Diego Fernando Martinez, Krishna Pattipati
          Ayala, Satnam Singh, and Pulak Bandyopadhyay. Event-driven data mining techniques
          for automotive fault diagnosis. In *Proc. of the 2010 Internat. Workshop on Principles
          of Diagnosis (DX 2010)*, 2010.

SKT05.   Miroslaw Staron, Ludwik Kuzniarz, and Christian Thurn. An empirical assessment
          of using stereotypes to improve reading techniques in software inspections. In *ACM
          SIGSOFT Software Engineering Notes*, volume 30, pages 1–7. ACM, 2005.

SKW04.   Miroslaw Staron, Ludwik Kuzniarz, and Ludwik Wallin. Case study on a process
          of industrial MDA realization: Determinants of effectiveness. *Nordic Journal of
          Computing*, 11(3):254–278, 2004.

SSBH14.   Giuseppe Scanniello, Miroslaw Staron, Håkan Burden, and Rogardt Heldal. On the
          effect of using SysML requirement diagrams to comprehend requirements: results from
          two controlled experiments. In *Proceedings of the 18th International Conference on
          Evaluation and Assessment in Software Engineering*, page 49. ACM, 2014.

Sta16.   Miroslaw Staron. Software complexity metrics in general and in the context of ISO
          26262 software verification requirements. In *Scandinavian Conference on Systems
          Safety*. http://gup.ub.gu.se/records/fulltext/233026/233026.pdf, 2016.

SW06.   Miroslaw Staron and Claes Wohlin. An industrial case study on the choice between
          language customization mechanisms. In *Product-Focused Software Process Improve-
          ment*, pages 177–191. Springer, 2006.

VF13.   Andreas Vogelsanag and Steffen Fuhrmann. Why feature dependencies challenge the
          requirements engineering of automotive systems: An empirical study. In *Requirements
          Engineering Conference (RE), 2013 21st IEEE International*, pages 267–272. IEEE,
          2013.

VS02.   Pablo Vidales and Frank Stajano.   The sentient car: Context-aware automotive
          telematics. In *Proceedings of the Fourth International Conference on Ubiquitous
          Computing*, pages 47–48, 2002.

Chapter 3
Automotive Software Development

Abstract In this chapter we describe and elaborate on software development processes in the automotive industry. We introduce the V-model for the entire vehicle development and we continue to introduce modern, agile software development methods for describing the ways of working of software development teams. We start by describing the beginning of all software development—requirements engineering—and we describe how requirements are perceived in automotive software development using text and different types of models. We discuss the specifics of automotive software development such as variant management, different integration stages of software development, testing strategies and the methods used for these. We review methods used in practice and explain how they should be used. We conclude the chapter with discussion on the need for standardization as the automotive software development is based on client-supplier relationships between the OEMs and the suppliers developing components of vehicles.

3.1 Introduction

Software development processes are at the heart of software engineering as they provide *structure and rigor* to the practices of developing software [C+90]. Software development processes consist of phases, activities and tasks which prescribe what actors should do. The actors can have different roles in software development such as software construction designers, software architects, project managers and quality managers.

The software development processes are organized in phases where the focus is on a specific part of software development. Historically these phases include:

1. requirements engineering—the phase where ideas about the functions of the software are created and broken down into requirements (atomic pieces of information about what should be implemented)
2. software analysis—the phase where the system analysis is conducted and high-level decisions about the allocation of functionality to the logical part of the system are made
3. software architecting—the phase where the software architects describe the high-level design of the software including its components and allocate them to computational nodes (ECUs)

© Springer International Publishing AG 2017
M. Staron, *Automotive Software Architectures*,
DOI 10.1007/978-3-319-58610-6_3

4. software design—the phase where each of the components is designed in detail
5. implementation—the phase where the design for each component is implemented in programming languages relevant for the design.
6. testing—the phase where the software is tested in a number of different ways, for example through unit and component tests.

These phases are often done in parallel as modern software development paradigms postulate that it is best to design, implement and test software iteratively. However, the prevalent software development model in the automotive industry is the so-called V-model where these phases are aligned to a V-shaped curve, where the design phases are on the left-hand side of the V and the testing phases are on the right-hand side of the V.

3.1.1 V-Model of Automotive Software Development

The V-model is illustrated in Fig. 3.1. This model is prescribed by international industry standards for development of safety-critical systems, like the ISO/IEC 26262 [ISO11].

In the figure, we also make a distinction between the responsibilities of OEMs (vehicle manufactures) and those of their suppliers. This distinction is important as it is often the phase where the handshaking between the suppliers and OEMs takes place, and therefore the requirements are used during the contract negotiations. In this context a detailed, unambiguous and correct requirements specification prevents potentially unnecessary costs related to the changes in requirements caused by misunderstandings between the OEMs and suppliers.

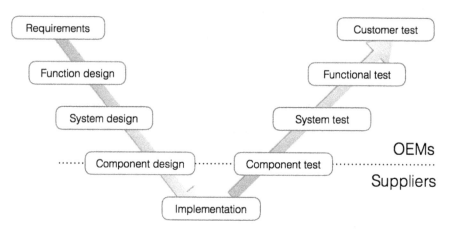

Fig. 3.1 V-shaped model of software development process in automotive software development

In the remainder of this chapter we go through the requirements engineering phase and the testing phase. The analysis and architecture phase are included in the next chapter while the detailed design phase is included in the latter part of the book.

3.2 Requirements

Requirements engineering is a discipline of vehicle development on the one hand and on the other hand a subdomain of software engineering and an initial phase of the software development lifecycle. It deals with the methods, tools and techniques for eliciting, specifying, documenting, prioritizing and quality assuring the requirements. The requirements themselves are very important for the quality of software in multiple senses as the quality is defined as *"The degree to which software fulfills the user requirements, implicit expectations and professional standards."* [C+90].

Requirements engineering in the automotive sector is increasingly about the software since the software is the source of the innovations. According to Houdek [Hou13] and a report about the innovation in the car industry [DB15], the number of functions in an average car grows much faster than the number of devices, with the number of systematic innovations growing faster than the individual innovations. The systematic innovations are systems of software functions rather than individual functions.

Therefore the discipline of requirements engineering is more about engineering than it is about innovation.

The length of an automotive requirements specification is in the range of 100,000 pages for a new car model according to Houdek, based on his study at Mercedes-Benz [Hou13], with ca. 400 documents of 250 pages each at the lowest specification level (component specifications), which are sent over to a large number of suppliers (usually over 100 suppliers, one for each ECU in the car).

Weber and Weisbrod [WW02] showed the complexity and size of requirements specifications in the automotive domain based on their experiences at Daimler-Chrysler. Their large software development projects can have as many as 160 engineers working on a single requirement specification and producing over 3 GB of requirements data. Weber and Weisbrod describe the process of requirements engineering in the following way: "Textual requirements are only part of the game – automotive development is too complex for text alone to manage." This quote reflects the state-of-the-practice of requirements engineering—that the requirements form only one part of the construction database. However, let's look at how the requirements are specified in the automotive domain. Similar challenges of linking requirements to other parts of the construction database can be also found in our previous studies in [MS08].

The requirements are often defined as *(1) A condition or capability needed by a user to solve a problem or achieve an objective. (2) A condition or capability that must be met or possessed by a system or system component to satisfy a contract, standard, specification, or other formally imposed documents. (3) A documented representation of a condition or capability as in (1) or (2)* [C$^+$90]. This definition stresses the link between the user of the system and the system itself, which is important for a number of reasons:

- Testability of the system—it should be clear how a requirement should be tested, e.g. what is the usage scenario realized by the requirement?
- Traceability of the functionality to design—it should be possible to trace which parts of the software realize the requirement in order to provide safety argumentation and enable impact/change management
- Traceability of the project progress—it should be possible to get an overview of which requirements have already been implemented and which are still to be implemented in the project

It is a very technical definition for something that is intuitively well known—a requirement is a way of communicating what we, the users, want in our dream car. In this sense it seems that the discipline of requirements engineering is simple. In practice, working with requirements is very complex as the ideas which we, users, have need to be translated to one of the millions of components of the car and its software. So, let's look at how the automotive companies work with our requirements or dreams.

We talk about software requirements engineering because the automotive industry has recognized the need to move innovation from the mechanical parts of the car to the electronics and software. The majority of us, the customers, buy cars today because they are fast (sporty), safe or comfortable. In many cases these properties are realized by adjusting the software that steers the parts of modern cars. For example we can have the same car with a software package that makes it extremely sporty—look at Tesla's "Insane" acceleration package or Volvo's Polestar performance package. These represent just two challenges which lead to two very important trends in automotive software requirements engineering:

1. Growing amount of software in contemporary cars—as the innovation is driven by software, the amount of software and its complexity grow exponentially. For example the amount of software in the 1990s was a few megabytes of binary code (e.g. Volvo S80) and today reaches over one gigabyte, excluding maps and other user data (e.g. Volvo XC90 of 2016).
2. Safety requirements posed by such standards as ISO 26262—as software steers more parts of the car, there is a larger probability that it can interfere with our driving and cause accidents and therefore it has to be safety-assured just like the software in airplanes and trains. The contemporary standard for functional safety (ISO/IEC 26262, Road vehicles—Functional safety) prescribes methods and processes to specify, design and verify/validate the software.

Automotive software requirements engineering therefore requires rigid processes for handling the construction of software for a car and therefore is very different from other types of software requirements engineering, such as for telecom or web design.

This chapter takes us through the theory of requirements engineering in automotive development by looking into two types of requirements—textual specifications and models used as requirements. It also helps us to explore the evolution of requirements engineering in automotive software development to finally draw on current trends and challenges for the future.

3.2.1 Types of Requirements in Automotive Software Development

When designing software for a car, the designers (who are often referred to as constructors) gradually break down the requirements from car level to component level. They also gradually refine them from textual requirements to models of behaviour of the software. This gradual refinement is due to the fact that the requirements have to be sent to Tier 1 suppliers for development and therefore should be as detailed as possible to enable their validation. In the automotive domain we have a number of tiers of suppliers:

- Tier 1—suppliers working directly with OEMs, usually delivering complete software and hardware subsystems and ECUs to the OEMs
- Tier 2—suppliers working with Tier 1 suppliers, delivering parts of the sub-products which are then delivered by Tier 1 suppliers to the OEMs; Tier 2 suppliers usually do not work directly with OEMs, which makes it even more important for the requirements to be detailed so that they can be correctly broken down by Tier 1 suppliers for Tier 2.
- Tier 3—suppliers working with Tier 2 suppliers, similarly to Tier 2 suppliers working with Tier 1 suppliers. Usually silicon vendors who deliver the hardware together with the drivers.

In this section we describe these different types of requirements, which can be found in these phases.

3.2.1.1 Textual Requirements

AUTOSAR is a great source of inspiration for research in automotive software development, and therefore let us look at the requirements in this standard—they are mostly textual. We use the same template as AUTOSAR for specifying requirements to provide an example of a requirement for keyless entry to the vehicle, as presented in Fig. 3.2.

The structure of the requirement is quite typical for requirements in general— it contains the description, the rationale and the use cases. So far we do not see anything specific. Nevertheless, if we look at the sheer size of such a specification—

REQ-1: Keyless vehicle entryt

Type	Valid
Description	It should be able to open the car with an RFID key
Rationale	The cars of our brand should all have the possibility to be opened using keyless solution. The majority of our competitors have an RFID sensors in the car that opens and starts the car based on the proximity of the designated driver who has the RFID sender (e.g.a card).
Use case	Keyless start-up
Dependencies	REQ-11: RFID implementation
Supporting material	---

Fig. 3.2 An example AUTOSAR requirement

over 1000 pages—we can see that we might confront issues; so let's discuss the kind of issues we can discover.

Rationale The textual requirements are used when describing high-level properties of cars. These types of requirements are mostly used in two phases—the requirements phase, when the specification of the car's functionality at a high level takes place, and at the component design phase, where large software requirements specification documents are sent to suppliers for development (although the textual requirements are often complemented by model-based requirements).

Method Specifying this kind of requirement rarely happens from scratch. Textual requirements are often specified based on models (e.g. UML domain models) and are intended to describe details of the inner workings of software systems. They are often linked to verification methods describing how the requirements should be verified—e.g. describing the test procedure for validating that the requirement is implemented correctly. Quite often it is the suppliers who do the verification, as many requirements demand specific test equipment to test their implementation. If this is the case, the OEMs choose a subset of requirements and verify them to check the correctness of the verification procedure on their side.

Format The text for the requirement is specified in the format which we can see in Fig. 3.2—tables with text. This format is very good if we can specify the details, but they are not very good when we want to communicate overviews and provide the context for the requirements. For that we need other types of requirements—use cases or models.

3.2.1.2 Use Cases

In software engineering the golden standard for specifying requirements is using use cases as defined by Jacobson, together with his Objectory methodology, in the

Fig. 3.3 An example use case specification with one use case

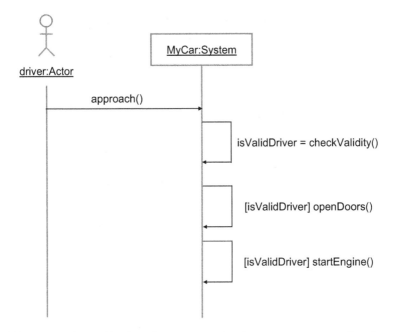

Fig. 3.4 An example specification of a use case using the message sequence charts/sequence diagrams

1990s [JBR97]. The use cases describe a course of interaction between an actor and the system under specification, for example as shown in Fig. 3.3, where the actor interacts with the car in the use case "Keyless start-up". The corresponding diagram (called the use case diagram in UML) is used to present which interactions (use cases) exist and how many actors are included in these interactions.

In the automotive industry this kind of requirements specification is the most common when describing the functions of the vehicles and their dependencies. It is used to describe how the actors (drivers or other cars) interact with the designed vehicle (the system) in order to realize a specific use case. This kind of specification is often described using the sequence diagrams of UML and we can see an example of such a specification in Fig. 3.4.

Rationale The use case specifications provide a high-level overview of the functionality of the designed system, such as a car, and therefore are very useful in the early phases of vehicle development. Usually these early phases are the functional design (use case diagrams) and the beginning of the system design (use case specifications).

Method Using the high-level descriptions of product properties, the functional designers break down these properties into usage scenarios. These usage scenarios provide a way to identify which of the functions (use cases) are of value to the customers and which are too cumbersome.

Format These kinds of specifications consist of three parts—(1) the use case diagram, (2) the use case specification using a sequence diagram, and (3) the textual specification of a use case detailing the steps of the interaction using somewhat structured natural language.

3.2.1.3 Model-Based Requirements

One method to provide more context to the requirements is to express them as models. This kind of representation can be done in two types of formalisms—UML-like models and Simulink models. In Fig. 3.5 we present an excerpt of a Simulink model for an ABS system from [Dem] and [RSB+13a].

The model shows how to implement the ABS system, but the most important property is that the model shows how the algorithm should behave and therefore how it should be verified.

Rationale Using models as requirements has been recognized by practitioners, and in an automotive software project up to 23% of models are used as requirements

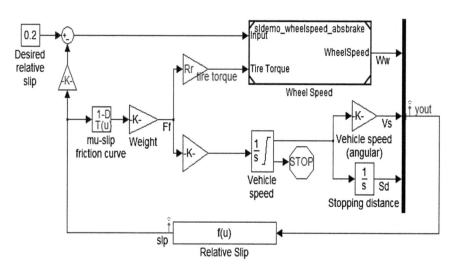

Fig. 3.5 An example Simulink model which can be used as a requirement to describe how to implement the ABS system

according to our previous studies [MS10b] and [MS10a]. According to the same studies, up to 13% of effort is spent in the software project to design these kinds of requirements.

Method The simulation models used for requirements engineering are often used as part of the process of system design and function design, where the software and system designers develop algorithms that describe how functions in modern cars are to be realized. These models can be automatically translated to C/C++ code using code generation, but it is rather uncommon. The reason is that these models describe entire functions which are often partitioned into different domains and spread over multiple components. Quite often these kinds of requirements are translated into textual specifications, shown in the previous subsection.

Format The models are expressed using Simulink or a variation of statechart such as Statemate or Petri nets. These simulation models detail the functions described in the use cases by adding the system view of the interaction—the blocks and signals. The blocks and signals represent the realization of the functionality in the car and are focused on one function only. These models are often used as specifications which are then detailed and often used to generate the source code automatically.

3.3   Variant Management

Having a good database of requirements and construction elements is key to success in automotive software engineering. This is dictated by the fact that the automotive market is based on *variability*—i.e. the locations in the product (software) where it can be configured. As customers we expect the ability to configure our car with the latest and greatest features of hardware, electronics and software.

There are two basic kinds of variability mechanisms in automotive software:

- Configuration—when we configure parameters of the software without modifying its internal structure. This kind of variability is often seen in the non-safety critical functions such as engine calibration or in configuring the availability of functions (e.g. rain sensors).
- Compilation—when we change the internal structure of the software, compile it and then deploy on the target ECU. This kind of variability is used when we need to ensure that the software always behaves in the same way, for example the availability of the function for collision avoidance by breaking.

In this section we explain the basics of these two mechanisms.

Fig. 3.6 Variability through configuration

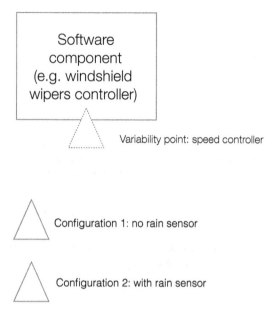

3.3.1 Configuration

Configuration is often referred to as runtime variability as changing the software can be done after the software is compiled. Figure 3.6 presents the conceptual view of this kind of variability.

In Fig. 3.6 we can see that we have one variant of the software component (rectangle) with one variability point (the dotted line triangle) which can be configured using two different configurations—without the rain sensor and with the rain sensor. This means that we compile the code for the software component once and then use two different configuration files when deploying the software.

The configuration as a variability mechanism has important implications for the designers of the software. The main implication is that the software has to be tested using multiple scenarios—i.e. the software designers need to be able to prevent use of the software component with invalid configurations.

3.3.2 Compilation

The compilation as a variability mechanism is fundamentally different as it results in a software component which cannot be modified (configured) after its compilation, during runtime. Therefore it is an example of so-called *design time variability* as the designers must decide during design which variant is being developed. This is

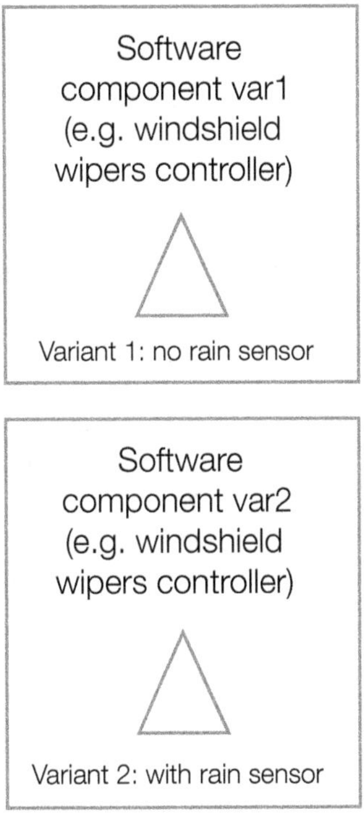

Fig. 3.7 Variability through
compilation

```
+-----------------------------------+
|             Software              |
|         component var1            |
|         (e.g. windshield          |
|         wipers controller)        |
|                                   |
|                /\                 |
|               /  \                |
|              /    \               |
|             /_____\              |
|                                   |
|     Variant 1: no rain sensor     |
+-----------------------------------+

+-----------------------------------+
|             Software              |
|         component var2            |
|         (e.g. windshield          |
|         wipers controller)        |
|                                   |
|                /\                 |
|               /  \                |
|              /    \               |
|             /_____\              |
|                                   |
|    Variant 2: with rain sensor    |
+-----------------------------------+
```

conceptually shown in Fig. 3.7 where we can see two different versions of the same component—with and without the rain sensor.

As Fig. 3.7 suggests, there are two different code bases for the software component—one with and one without the rain sensor. This means that the development of these two variants can be decoupled from each other, but that also means that the designers have to maintain two different code bases at the same time. This parallel maintenance means that if there are defects in the common code then both code bases need to be updated and tested.

The main advantage of this kind of variability mechanism is the assurance that the code is not tampered with in any way after the compilation. The code can be tested, and once deployed there is no way that an incorrect configuration can break the quality of the component. However, the main disadvantage of this type of variability management mechanism is the high cost of maintenance of the code base—parallel maintenance.

3.3.3   Practical Variability Management

Both of the above variability management mechanisms are used in practice. Compile time variability is used when the software is an integral part of an ECU whereas configuration is used when the software can be calibrated to different types of configurations during deployment (e.g. configuration on the assembly line, calibration of the engine and gearbox depending on the powertrain performance settings).

3.4 Integration Stages of Software Development

On the left-hand side of the V-model the main type of activity is refinement of requirements in multiple ways. On the right-hand side of the model the main activity type is integration followed by testing.

In short, integration is the activity where software construction engineers integrate their code with the code of other components and with the hardware. In the first integration stages the hardware is usually simulated hardware in order to allow for unit and component testing (described in Sect. 3.5). In the later integration phases the software code is integrated together with the target hardware, which is then integrated into a complete electrical/electronic system of the car (Table 3.1).

Figure 3.8 shows an example software integration of software modules and components into an electrical system. What is important to notice is the fact that the integration steps (vertical solid black lines) are not synchronized as the development of different software modules is done at different pace.

In practice this figure is even more complicated, as the integration plan is often a document with several dimensions. Each integration cycle (which is what we show in Fig. 3.4) is done several times during the project. First, the so-called basic

Table 3.1 Types of integration

Type	Description
Software integration	This type of integration means that two (or more) software components are put together and their joint functionality is tested. The usual means of integration depend on the what is integrated—it can be merging of the source code if the integration is on the source code level; it can be linking of two binary code bases together; or it can be parallel execution to test interoperability. The main testing techniques are unit and component testing, described in Sect. 3.5
Software-hardware integration	This type of integration means that the software is integrated (deployed) to the target hardware platform. In this type of integration, the focus is on the ability of the complete ECU to be executed and the main testing type is component testing, described in Sect. 3.5
Hardware integration	This type of integration means that the focus is on the integration of the ECUs with the electrical system. In this type of integration the focus is on the interoperability of the nodes and basic functionality, such as communication. The testing related to this type of integration is system testing

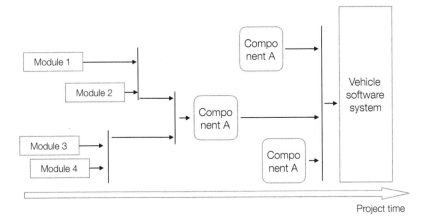

Fig. 3.8 Software integration with integration steps

software is integrated (functionality like the boot code, and communication) and then higher level functionality is added, according to the functional architecture as described in Chap. 2.

3.5 Testing Strategies

Requirements engineering progresses from higher abstraction levels towards more detailed, lower abstraction levels. Testing is the opposite. When the testers test the software they start from the most atomic type of testing—unit testing—where they test each function and each line of code. Then they gradually progress by testing entire components (i.e. multiple units linked together), then the entire system and finally each function. Figure 3.9 shows the right-hand side of the V-model with a focus on the testing phases.

In the coming subsections we look deeper into the testing phases of the automotive software.

3.5.1 Unit Testing

Unit test is the basic test, which is performed on individual entities of software such as classes, source code modules and functions. The goal of unit testing is to find defects related to the implementation of atomic functions/methods in the source code.

The basic scheme of unit testing is the creation of automated test cases which combine individual methods with the data that is needed to achieve the needed

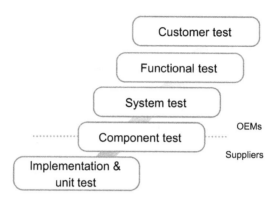

Fig. 3.9 Testing phases in automotive software development

```
 1  using System;
 2  using Microsoft.VisualStudio.TestTools.UnitTesting;
 3  using WindshieldSimulator;
 4
 5  namespace WindshieldTest
 6  {
 7      [TestClass]
 8      public class BasicSuite
 9      {
10          // unit test method
11          [TestMethod]
12          public void TestCreationInitialState()
13          {
14              // arrange
15              WindshieldWiper pWiper;
16
17              // act
18              pWiper = new WindshieldWiper();
19
20              // assert
21              Assert.AreEqual(pWiper.Status,
22                              position.closed,
23                              "Initial status should be /closed/");
24          }
25      }
26  }
```

Fig. 3.10 Example unit test for testing the status of windshield wiper module

quality. The result is then compared to the expected result, usually with the help of assertions. An example of a unit test is presented in Fig. 3.10.

The unit test presented in Fig. 3.10 is a test for correctness of the creation of object "WindshieldWiper"—a unit under test (UAT). This particular test code is written in C# and in practice the test code can be written in almost any programming language. The principles, however, are the same for all unit tests.

Of interest for our chapter are lines 14–23, as they contain the actual test code. Line 15 is the arrangement line which prepares (sets up) the test case—in our example it declares a variable which will be assigned to the object of the class WindshieldWiper. Line 18 is the actuation line which executes the actual test code—in our example creates the object of the WindshieldWiper class.

The most interesting are lines 21–23 since they contain the so-called assertion. The assertion is a condition which should be fulfilled after the execution of the test code. In our example the assertion is that the status of the newly created object (line 21) is "closed" (line 22). If it is not the case, then the error message is logged in the testing environment (line 32) and the execution of the new test cases continues.

Unit testing is often perceived as the simplest type of testing and is most often automated. Frameworks like CppUnit, JUnit or Google test framework can orchestrate the execution of unit tests, allowing us to quickly execute the entire set of tests (called test suites) without the need for manual intervention.

Automated unit tests are also reused in several ways, for example to create nightly regression test suites or to create the so-called "smoke testing" where testers randomly execute test cases to check whether the system exposes random behavior.

It is also important to notice that reuse of test cases needs to be accompanied by the methods to prioritize test cases, e.g. by identifying risky areas in source code [ASM+14] or focusing on code that was changed since the last test run [KSM+15, SHF+13]. It is also important to trace the test process in the context of software reliability growth [RSM+13, RSB+13b].

We can also see that if the test case finds a problem (fails), then troubleshooting is relatively simple—we know which code was executed and under which conditions. This knowledge allows the testers to quickly describe where the defect is or even suggest how to fix it.

3.5.2 Component Testing

This is sometimes also called *integration testing*, as the goal of this type of testing is to test the integrations, i.e. links, between units of code within one of many components. The main characteristic which differentiates component tests from unit tests is that in component testing we use stubs to simulate the environment of the tested component or the group of the components. This is illustrated in Fig. 3.11.

In contrast to unit tests, component tests focus on the interaction between the stubs and the component under test. The goal of this type of testing is to verify that the structure and behavior of the interfaces is implemented correctly.

We should also mention that the number of stubs in the system decreases as the development project progresses. With the progress of the development, new components are designed and they replace the stubs. Hence the nickname of this type of testing—"integration testing".

In automotive systems this type of testing is often done by simulating the environment using either models (the so-called Model-In-the-Loop or MIL testing)

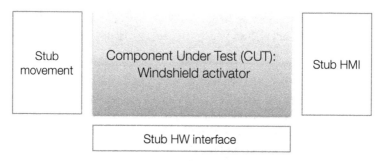

Fig. 3.11 Component under test with the simulated environment

or hardware simulators (the so-called Hardware-In-the-Loop or HIL testing). An example of equipment for HIL testing is presented in Fig. 3.12.

Figure 3.12 shows a testing rig from dSpace, which is widely used in the automotive industry to test components by simulating the external environment.

Since the environment of the components is simulated, the non-functional properties of the components are often hard to test or require very detailed simulations. The very detailed simulations, however, also tend to be very costly.

3.5.3   System Testing

System testing is the phase of testing when the entire system is assembled and tested as a whole. The focus of system testing is on checking whether the system fulfills its specifications in a number of ways. The system testing focuses on verifying the following aspects:

1. functionality—testing whether the system has the functionality as specified in the requirements specification
2. interoperability—testing whether the system can connect to the other systems which are designed to interact with the system under test
3. performance—testing whether the system under test performs within the specified limits (e.g. timing limits, capacity limits)
4. scalability—testing whether the system's operation scales up and down (e.g. whether the communication buses operate with 80 and 120 ECUs connected)
5. stress—testing whether the system operates correctly under high load (e.g. when the maximum capacity of the communication buses is reached)
6. reliability—testing whether the system operates correctly during a specific period of time
7. regulatory and compliance—testing whether the system fulfills legal and regulatory requirements (e.g. carbon dioxide emissions)

Fig. 3.12 HIL testing
rig—Image source: dSPACE
GmbH. Copyright 2015
dSPACE GmbH—reprinted
with permission

System testing is usually the first testing phase when the above aspects can be tested and therefore it is usually the most effective way of testing. However, it is also very costly way of testing and very inefficient, as fixing the defects found in this phase requires frequent changes in multiple components.

In the automotive software this type of testing is often done using the so-called "box cars"—the entire electrical system being set up on tables without the chassi and the hardware components.

Test ID	T0001
Description	Test of the basic function of windshield wipers. The goal of the test is to verify that the windshield wipers can make one sweep with the engine turned off.

Action/Step	Expected result
Start ignition	Battery icon on the dashboard lit red; windshield wipers are in the position "closed".
Push the windshield wipers level one step forward	The windshield wipers start to move to the position "open"
Wait 20 seconds	The windshield wipers go back to the position "closed"
Turn off ignition	All icons on the car's dashboard turn off; windshield wipers are in position "closed".

Fig. 3.13 Example of a functional test

3.5.4 Functional Testing

The functional testing phase focuses on verifying that the functions of the system work according to their specification. They correspond to the functional requirements in the form of use cases and are quite often specified according to the use cases. Figure 3.13 presents an example of a functional test—specified as a table.

What is important in this example is the specification, which is similar to the specification of a use case—the description of the action (step) on the left-hand side together with the expected outcome on the right-hand side. We can also observe that the functional test does not require the knowledge of the actual construction of the system under test (SUT), which led to the nickname of these tests as "black-box testing".

We should not focus on the simplicity of the example because functional testing is often the most effort-intensive type of testing. It is often done in a manual manner and requires sophisticated equipment to conduct.

Examples of sophisticated functional test cases are safety test cases where OEMs test their safety systems. To be able to test such a function, car manufacturers need to recreate the situation where the system could be activated and check whether it was activated. They also need to recreate the situation when it should not be activated and test that it was not activated.

When the functional test fails, it is rather difficult to find the defect, as the number of construction elements which take part in the interaction can be quite large—in our example the failure of the functional test case could be caused by anything from mechanical failure of the battery to design defect in the software. Therefore functional testing is often used after the other tests are conducted to validate functionality rather than to verify the design.

3.5.5   *Pragmatics of Testing Large Software Systems: Iterative Testing*

As the electrical system of contemporary cars is very complex, OEMs often apply concepts of interactive testing to their development. Concept of iterative testing means that the functionality of the software is divided into levels (as prescribed by the functional architecture described in Chap. 2) and the functions are tested using unit, component, system and functional testing per layer. This means that the basic functionality such as booting up of the electronics, starting up of the communication protocols, running diagnostics, etc. are tested first and the more advanced functions such as lighting, steering, and braking are tested later, followed by more advanced functions such as driver alerts.

3.6   Construction Database and Its Role in Automotive Software Engineering

All these types of requirements need to come together somehow and that's why we have the process and the infrastructure for requirements engineering. Let us start with the infrastructure—usually named the design or construction database. In the light of work of Weber and Weisbrod [WW02] it is called the common information model. Figure 3.14 shows how this design database is used. The construction

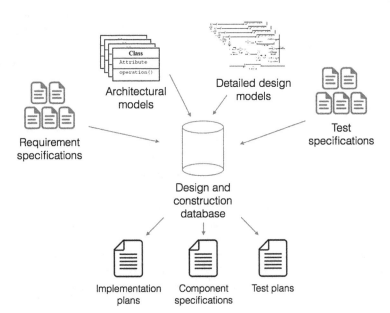

Fig. 3.14 Design database

database contains all elements of the design of the electrical system of the vehicle—components, electronic control units, systems, controllers, etc. The structure of such a database is hierarchical and reflects the structure of the vehicle. Each of the elements in the database has a set of requirements linked to it. The requirements are also linked to one another to show how they are broken down. Such a database grows over time and is version-controlled as different versions of the same elements can be used in different vehicles (e.g. different year models of the same car or different cars).

An example of such a system is described by Chen et al. [CTS+06] and has been developed by the company Systemite, which specializes in databases for vehicle construction. Such a database structures all the elements of the construction of the electronics of the vehicle and links all artifacts to the construction elements. An example of a construction element is the engine's electronic control unit, and all the functions that use this control unit are linked to it.

Such a database usually has a number of views which show the required set of details—functional view, architectural view, topological view and software components' view. Each view provides the corresponding entry point and shows the relevant elements, but the database is always in a consistent state where all the links are valid.

The database is used to generate construction specifications for different actors. For each supplier that delivers an ECU, the database generate the set of all requirements which are linked to the ECU and all models which describe the behaviour of the ECU. Sometimes, depending on the situation, the documentation contains even the simulation models for the functions which are to be included in the ECU.

One of the commercial tools available on the market which is used as a construction database is the tool SystemWeaver provided by Systemite. The main strength of such a tool is the ability to link all elements together. In Fig. 3.15 we can see how the requirements are linked to the software architecture model. On the left-hand side we can see that the requirements are part of an element (e.g. "Adjust speed" as part of the "Adaptive cruise control"), and on the right-hand side another requirement visualized as a diagram.

Such tools provide specific views, for example listing all requirements linked to a specific function as shown in Fig. 3.16. As part of that view we can see that the text is complemented with figures which allow the analysts to be more specific when specifying requirements and allow the designers to understand the requirements better.

The ability to link the elements from different views (e.g. requirements and components) and provide a graphical overview of these elements allows the architects to quickly perform change impact analyses and reason about their architectural choices. Such a dynamic creation of views is very important when assessing architectures (e.g. during ATAM assessments). An example of such a view is one showing a set of architectural components used in realization of a specific user function, as shown in Fig. 3.17.

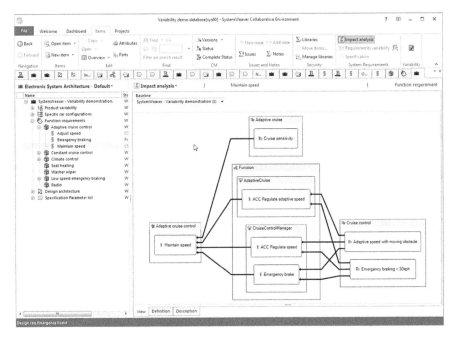

Fig. 3.15 Design database linking requirements to architectural elements. Copyright 2016, Systemite—reprinted with permission

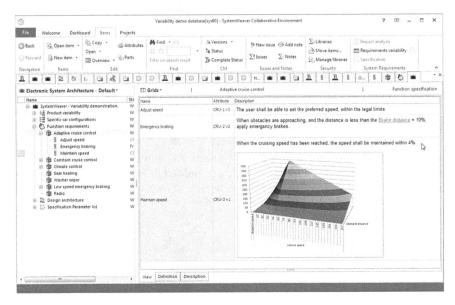

Fig. 3.16 Design database listing requirements for a specific function. Copyright 2016, Systemite—reprinted with permission

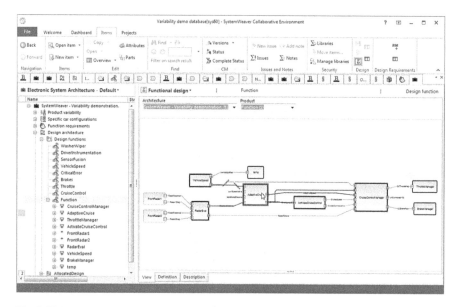

Fig. 3.17 Design database showing architectural components used when designing a specific function. Copyright 2016, Systemite—reprinted with permission

The system construction database can also help us in linking requirements to test cases during the test planning phase—as shown in Fig. 3.18.

It can also assist us in tracking the progress of testing—Fig. 3.19. Since the number of requirements is so large in automotive systems, tracking the progress of whether they are tested is also not trivial. Therefore a unified view is needed where the project can track the test cases that are planned to cover certain requirements, as well as those that they were executed and what the result of the execution was.

The construction database and modelling tool provide the project teams with a consistent view on their software system. In the case of software architectures this tool allows us to link together all the views presented in Chap. 2 (such as physical, logical, and deployment) and therefore avoid unnecessary work to keep documents in a steady and consistent state. Most of the tools available for this purpose provide the possibility to handle multiple parallel versions and baselines, which is essential in the development of automotive software.

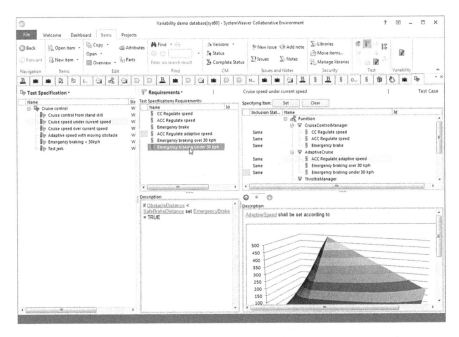

Fig. 3.18 Linking test cases to requirements. Copyright 2016, Systemite—reprinted with permission

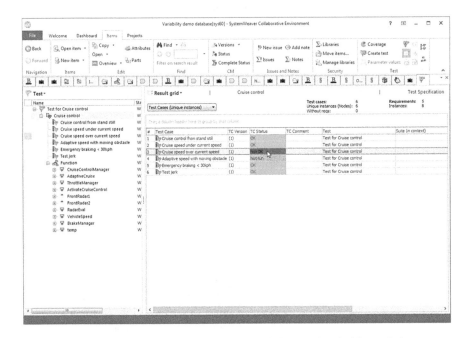

Fig. 3.19 Tracking test progress. Copyright 2016, Systemite—reprinted with permission

3.7 Further Reading

In this chapter we outlined the practical aspects of automotive software development from a bird's eye perspective. Interested readers can turn to more literature in the area to dive deeper into details.

For the automotive software processes we recommend the book by Schäuffele and Zurawka [SZ05], which presents a classical view on automotive software development, starting from low-level processor programming and moving on to advanced functionality development.

The classical paper by Broy [Bro06] describing the challenges in automotive software engineering is the next step to understanding the dynamics of automotive software engineering in general. This reading can be complemented by the paper by Pretschner et al. [PBKS07], where the focus is on the future of automotive software development.

Readers interested in the management of variability in general should explore the work of Bosch et al. [VGBS01, SVGB05] or [BFG+01]. The work is based on software product lines, but applies very well to the automotive sector. This can be complemented with more recent developments in this area—software ecosystems and their realization in the automotive sector [EG13, EB14].

Otto et al. [Ott12] and [Ott13] presents a study on requirements engineering at Mercedes-Benz, where they classified over 5800 requirement review protocols to their quality model. Their results showed that textual requirements (or natural language requirements as they are called in the publication) are prone to such problems as inconsistency, incompleteness and ambiguity—with about 70% of defects in requirements falling into these categories. In the light of this article we can see the need for complementing the textual requirements with more context, provided by use case models, user stories and use cases.

Törner et al. [TIPÖ06] presented a similar study but of the requirements at Volvo Car Group. In contrast to the study of Otto et al. [Ott12], these authors studied the use case specifications and not the textual requirements. The results, however, are similar, as the main types of defects are missing elements (correctness in Otto et al.'s model) and incorrect linguistics (ambiguity in Otto et al.'s model).

Eliasson et al. [EHKP15] described further experiences from Volvo Car Group where they explored challenges with requirements engineering at large in a mechatronics development organization. Their findings showed that there is a lot of communication in parallel to the requirements specification. The stakeholders in the requirements specification frequently mentioned the need to have a good network in order to specify the requirements correctly. This indicates the challenges described previously in this chapter that the requirements need more context than is usually provided in just the specification (especially the textual specification).

Mahally et al. [MMSB15] identified requirements to be the main barriers and enablers in moving towards Agile mechatronics organizations. Although today OEMs try to move towards fast development of mechatronics and reduce the cycle time by using Agile software development approaches, the challenges are that we do not know upfront whether a requirement requires the development of electronics or is only a software requirement. According to Mahally et al. that kind of problem needs to be solved, and based on the prediction of Houdek [Hou13] this kind of issue might be coming to an end as device development flattens out and most of the requirements become software requirements. Similar challenges were presented by Pernstål et al. [PGFF13] who found that requirements engineering is one of the top improvement areas for automotive OEMs. The ability to communicate via requirements was also an important part.

At Audi, Allmann et al. [AWK$^+$06] presented the challenges in the requirements communication on the boundary between the OEMs and their suppliers. They identified the need for better communication and the challenges of communicating through textual representations. They recognized the need for tighter partnerships as there is an inherent deficiency in communicating through requirements—transferring knowledge through an intermediate medium. Therefore they recommended integrating systems to minimize knowledge loss via transfer of documents.

Siegl et al. [SRH15] presented a method for formalizing requirements specifications using the Time Usage Model and applied it successfully to a requirements specification from one of the German OEMs. The evaluation study showed an increase in test coverage and increased quality of the requirements specification.

At BMW, Hardt et al. [HMB02] demonstrated the use of formalized domain engineering models in order to reason about the dependencies between requirements in the presence of variants. Their approach provided a simplistic, yet powerful, formalism and its strength was industrial applicability.

A study of the functional architecture of a car project at BMW and the requirements linked to the functions by Vogelsang and Fuhrmann [VF13] showed that 85% of functions are dependent on one another and that these dependencies cause a significant number of problems in software projects. This study shows the complexity of the functional decomposition of the vehicle's design and the complexity of its description.

At Bosch, the longitudinal study of a 5-year project by Langenfeld et al. [LPP16] showed that 61% of defects in requirements come from the incompleteness or incorrectness of the requirements specifications.

One of interesting trends in requirements engineering is the automatization of tasks of requirements engineers. One of such tasks is the discovery of non-functional requirements. This task is based on reading the specifications of functional requirements and identifying phrases which should transform into non-functional requirements. A study on the automation of this task has been conducted by Cleland-Huang et al. [CHSZS07]. The study showed that the automated classification of requirements could be as good as 90%, but at this stage cannot replace the manual classifiers.

3.7.1 Requirements Specification Languages

A model for requirements traceability [DPFL10] DARWIN4Req has been proposed
to address the challenges related to the ability to follow the requirements' lifecycle.
The model allows us to link requirements expressed in different formalities (e.g.
UML, SySML) and connect them to one another. However, to the best of our
knowledge, the model and the tool have not been adopted on a wider scale yet.

EAST-ADL [DSLT05] is an architecture specification language which contains
elements to capture requirements and link them to the architectural design. The
approach is similar to that of SySML but with the difference that there is no
dedicated requirements specification diagram. EAST-ADL has been demonstrated
to work in industry; however, it is not a standard for automotive OEMs yet.
Mahmud [MSL15] presented a language ReSA that complements the EAST-ADL
modelling language with the possibility to analyze and validate requirements (e.g.
basic consistency checks).

For non-functional requirements in the domain of safety, Peraldi [PFA10] has
proposed another extension of the EAST-ADL language which allows for increased
traceability of requirements and their linking to non-functional properties of the
designed embedded software (e.g. Safety).

Mellegård and Staron [MS09] and [MS10c] conducted an empirical study on
the impact of using hierarchical graphical requirements specification on the quality
of change impact assessment. For this purpose they designed a requirements'
specification language based on the existing formalism—Requirements Abstraction
Model. The results showed that the graphical overview of the dependencies between
requirements introduces significant improvement [KS02].

Finally, the use of models as core artifacts in software development in the
automotive domain has been studied in the context of MDA (Model-Driven
Architecture) [SKW04a, SKW04b, SKW04c]. The important aspect is the evolution
of models throughout the lifecycle.

3.8 Summary

Correct, unambiguous and consistent requirements specifications are foundations
for high-quality software in general and in the automotive embedded systems in
particular. In this chapter we introduced the most common types of requirements
used in this domain and provided their main strengths.

Based on the current state of evolution of automotive software we could observe
three trends in requirements engineering for automotive embedded systems—
(1) agility in requirements specification, (2) increased focus on non-functional
requirements and (3) increased focus on security as a domain for requirements.
Towards the end of the chapter we also provided an overview of the requirements
practices at some of the vehicle manufacturers (Mercedes Benz, Audi, BMW and

Volvo) based on documented experiences at these companies. We have also pointed out a number of directions for further reading for the interested.

In our future work we plan to review the requirements engineering practices in the main automotive OEMs and identify their differences and commonalities.

References

ASM+14. Vard Antinyan, Miroslaw Staron, Wilhelm Meding, Per Österström, Erik Wikstrom, Johan Wranker, Anders Henriksson, and Jörgen Hansson. Identifying risky areas of software code in agile/lean software development: An industrial experience report. In *Software Maintenance, Reengineering and Reverse Engineering (CSMR-WCRE), 2014 Software Evolution Week-IEEE Conference on*, pages 154–163. IEEE, 2014.

AWK+06. Christian Allmann, Lydia Winkler, Thorsten Kölzow, et al. The requirements engineering gap in the oem-supplier relationship. *Journal of Universal Knowledge Management*, 1(2):103–111, 2006.

BFG+01. Jan Bosch, Gert Florijn, Danny Greefhorst, Juha Kuusela, J Henk Obbink, and Klaus Pohl. Variability issues in software product lines. In *International Workshop on Software Product-Family Engineering*, pages 13–21. Springer, 2001.

Bro06. Manfred Broy. Challenges in automotive software engineering. In *Proceedings of the 28th international conference on Software engineering*, pages 33–42. ACM, 2006.

C+90. IEEE Standards Coordinating Committee et al. IEEE Standard glossary of software engineering terminology (IEEE Std 610.12-1990). los alamitos. *CA: IEEE Computer Society*, 1990.

CHSZS07. Jane Cleland-Huang, Raffaella Settimi, Xuchang Zou, and Peter Solc. Automated classification of non-functional requirements. *Requirements Engineering*, 12(2):103–120, 2007.

CTS+06. DeJiu Chen, Martin Törngren, Jianlin Shi, Sebastien Gerard, Henrik Lönn, David Servat, Mikael Strömberg, and Karl-Erik Årzen. Model integration in the development of embedded control systems-a characterization of current research efforts. In *Computer Aided Control System Design, 2006 IEEE International Conference on Control Applications, 2006 IEEE International Symposium on Intelligent Control, 2006 IEEE*, pages 1187–1193. IEEE, 2006.

DB15. Jan Dannenberg and Jan Burgard. 2015 car innovation: A comprehensive study on innovation in the automotive industry. 2015.

Dem. Simulink Demo. Modeling an anti-lock braking system.

DPFL10. Hubert Dubois, Marie-Agnès Peraldi-Frati, and Fadoi Lakhal. A model for requirements traceability in a heterogeneous model-based design process: Application to automotive embedded systems. In *Engineering of Complex Computer Systems (ICECCS), 2010 15th IEEE International Conference on*, pages 233–242. IEEE, 2010.

DSLT05. Vincent Debruyne, Françoise Simonot-Lion, and Yvon Trinquet. EAST–ADL – An architecture description language. In *Architecture Description Languages*, pages 181–195. Springer, 2005.

EB14. Ulrik Eklund and Jan Bosch. Architecture for embedded open software ecosystems. *Journal of Systems and Software*, 92:128–142, 2014.

EG13. Ulrik Eklund and Håkan Gustavsson. Architecting automotive product lines: Industrial practice. *Science of Computer Programming*, 78(12):2347–2359, 2013.

EHKP15. Ulf Eliasson, Rogardt Heldal, Eric Knauss, and Patrizio Pelliccione. The need of complementing plan-driven requirements engineering with emerging communication: Experiences from Volvo Car Group. In *Requirements Engineering Conference (RE), 2015 IEEE 23rd International*, pages 372–381. IEEE, 2015.

HMB02.   Markus Hardt, Rainer Mackenthun, and Jürgen Bielefeld.   Integrating ECUs in
           vehicles-requirements engineering in series development. In *Requirements Engineer-
           ing, 2002. Proceedings. IEEE Joint International Conference on*, pages 227–236.
           IEEE, 2002.
Hou13.   Frank Houdek.   Managing large scale specification projects.   In *Requirements
           Engineering foundations for software quality, REFSQ*, 2013.
ISO11.   ISO. 26262–road vehicles-functional safety. *International Standard ISO*, 26262, 2011.
JBR97.   Ivar Jacobson, Grady Booch, and Jim Rumbaugh. The objectory software development
           process. *ISBN: 0-201-57169-2, Addison Wesley*, 1997.
KS02.   Ludwik Kuzniarz and Miroslaw Staron. On practical usage of stereotypes in UML-
           based software development. *the Proceedings of Forum on Design and Specification
           Languages, Marseille*, 2002.
KSM$^+$15. Eric Knauss, Miroslaw Staron, Wilhelm Meding, Ola Söder, Agneta Nilsson, and
           Magnus Castell. Supporting continuous integration by code-churn based test selection.
           In *Proceedings of the Second International Workshop on Rapid Continuous Software
           Engineering*, pages 19–25. IEEE Press, 2015.
LPP16.   Vincent Langenfeld, Amalinda Post, and Andreas Podelski. Requirements Defects over
           a Project Lifetime: An Empirical Analysis of Defect Data from a 5-Year Automotive
           Project at Bosch. In *Requirements Engineering: Foundation for Software Quality*,
           pages 145–160. Springer, 2016.
MMSB15.   Mahshad M Mahally, Miroslaw Staron, and Jan Bosch.   Barriers and enablers for
           shortening software development lead-time in mechatronics organizations: A case
           study. In *Proceedings of the 2015 10th Joint Meeting on Foundations of Software
           Engineering*, pages 1006–1009. ACM, 2015.
MS08.   Niklas Mellegård and Miroslaw Staron. Methodology for requirements engineering
           in model-based projects for reactive automotive software. In *18th ECOOP Doctoral
           Symposium and PhD Student Workshop*, page 23, 2008.
MS09.   Niklas Mellegård and Miroslaw Staron. A domain specific modelling language for
           specifying and visualizing requirements. In *The First International Workshop on
           Domain Engineering, DE@ CAiSE, Amsterdam*, 2009.
MS10a.   Niklas Mellegård and Miroslaw Staron. Characterizing model usage in embedded
           software engineering: a case study. In *Proceedings of the Fourth European Conference
           on Software Architecture: Companion Volume*, pages 245–252. ACM, 2010.
MS10b.   Niklas Mellegård and Miroslaw Staron.   Distribution of effort among software
           development artefacts: An initial case study. In *Enterprise, Business-Process and
           Information Systems Modeling*, pages 234–246. Springer, 2010.
MS10c.   Niklas Mellegård and Miroslaw Staron. Improving efficiency of change impact
           assessment using graphical requirement specifications: An experiment. In *Product-
           focused software process improvement*, pages 336–350. Springer, 2010.
MSL15.   Nesredin Mahmud, Cristina Seceleanu, and Oscar Ljungkrantz. ReSA: An ontology-
           based requirement specification language tailored to automotive systems. In *Industrial
           Embedded Systems (SIES), 2015 10th IEEE International Symposium on*, pages 1–10.
           IEEE, 2015.
Ott12.   Daniel Ott. Defects in natural language requirement specifications at Mercedes-Benz:
           An investigation using a combination of legacy data and expert opinion. In *Require-
           ments Engineering Conference (RE), 2012 20th IEEE International*, pages 291–296.
           IEEE, 2012.
Ott13.   Daniel Ott. Automatic requirement categorization of large natural language specifi-
           cations at Mercedes-Benz for review improvements. In *Requirements Engineering:
           Foundation for Software Quality*, pages 50–64. Springer, 2013.
PBKS07.   Alexander Pretschner, Manfred Broy, Ingolf H Kruger, and Thomas Stauner. Software
           engineering for automotive systems: A roadmap.   In *2007 Future of Software
           Engineering*, pages 55–71. IEEE Computer Society, 2007.

PFA10.   Marie-Agnès Peraldi-Frati and Arnaud Albinet. Requirement traceability in safety crit-
            ical systems. In *Proceedings of the 1st Workshop on Critical Automotive applications:
            Robustness & Safety*, pages 11–14. ACM, 2010.

PGFF13.   Joakim Pernstå, Tony Gorschek, Robert Feldt, and Dan Florén. Software process
            improvement in inter-departmental development of software-intensive automotive
            systems – A case study. In *Product-Focused Software Process Improvement*, pages 93–
            107. Springer, 2013.

RSB+13a.   Rakesh Rana, Miroslaw Staron, Christian Berger, Jörgen Hansson, Martin Nilsson, and
            Fredrik Törner. Improving fault injection in automotive model based development
            using fault bypass modeling. In *GI-Jahrestagung*, pages 2577–2591, 2013.

RSB+13b.   Rakesh Rana, Miroslaw Staron, Claire Berger, Jorgen Hansson, Martin Nilsson, and
            Fredrik Torner. Evaluating long-term predictive power of standard reliability growth
            models on automotive systems. In *Software Reliability Engineering (ISSRE), 2013
            IEEE 24th International Symposium on*, pages 228–237. IEEE, 2013.

RSM+13.   Rakesh Rana, Miroslaw Staron, Niklas Mellegård, Christian Berger, Jörgen Hansson,
            Martin Nilsson, and Fredrik Törner. Evaluation of standard reliability growth models
            in the context of automotive software systems. In *Product-Focused Software Process
            Improvement*, pages 324–329. Springer, 2013.

SHF+13.   Miroslaw Staron, Jorgen Hansson, Robert Feldt, Anders Henriksson, Wilhelm Meding,
            Sven Nilsson, and Christoffer Hoglund. Measuring and visualizing code stability – A
            case study at three companies. In *Software Measurement and the 2013 Eighth Interna-
            tional Conference on Software Process and Product Measurement (IWSM-MENSURA),
            2013 Joint Conference of the 23rd International Workshop on*, pages 191–200. IEEE,
            2013.

SKW04a.   Miroslaw Staron, Ludwik Kuzniarz, and Ludwik Wallin. Case study on a process
            of industrial MDA realization: Determinants of effectiveness. *Nordic Journal of
            Computing*, 11(3):254–278, 2004.

SKW04b.   Miroslaw Staron, Ludwik Kuzniarz, and Ludwik Wallin. A case study on industrial
            MDA realization – Determinants of effectiveness. *Nordic Journal of Computing*,
            11(3):254–278, 2004.

SKW04c.   Miroslaw Staron, Ludwik Kuzniarz, and Ludwik Wallin. Factors determining effective
            realization of MDA in industry. In K. Koskimies, L. Kuzniarz, Johan Lilius, and Ivan
            Porres, editors, *2nd Nordic Workshop on the Unified Modeling Language*, volume 35,
            pages 79–91. Abo Akademi, 2004.

SRH15.   Sebastian Siegl, Martin Russer, and Kai-Steffen Hielscher. Partitioning the require-
            ments of embedded systems by input/output dependency analysis for compositional
            creation of parallel test models. In *Systems Conference (SysCon), 2015 9th Annual
            IEEE International*, pages 96–102. IEEE, 2015.

SVGB05.   Mikael Svahnberg, Jilles Van Gurp, and Jan Bosch. A taxonomy of variability
            realization techniques. *Software: Practice and Experience*, 35(8):705–754, 2005.

SZ05.   Jörg Schäuffele and Thomas Zurawka. *Automotive software engineering – Principles,
            processes, methods and tools*. 2005.

TIPÖ06.   Fredrik Törner, Martin Ivarsson, Fredrik Pettersson, and Peter Öhman. Defects in
            automotive use cases. In *Proceedings of the 2006 ACM/IEEE international symposium
            on Empirical software engineering*, pages 115–123. ACM, 2006.

VF13.   Andreas Vogelsanag and Steffen Fuhrmann. Why feature dependencies challenge the
            requirements engineering of automotive systems: An empirical study. In *Requirements
            Engineering Conference (RE), 2013 21st IEEE International*, pages 267–272. IEEE,
            2013.

VGBS01.   Jilles Van Gurp, Jan Bosch, and Mikael Svahnberg. On the notion of variability
            in software product lines. In *Software Architecture, 2001. Proceedings. Working
            IEEE/IFIP Conference on*, pages 45–54. IEEE, 2001.

WW02.   Matthias Weber and Joachim Weisbrod. Requirements engineering in automotive
            development-experiences and challenges. In *Requirements Engineering, 2002. Pro-
            ceedings. IEEE Joint International Conference on*, pages 331–340. IEEE, 2002.

Chapter 4
AUTOSAR Standard

Darko Durisic
Volvo Car Group, Gothenburg, Sweden

Abstract In this chapter, we describe the role of the AUTOSAR (AUTomotive Open System ARchitecture) standard in the development of automotive system architectures. AUTOSAR defines the reference architecture and methodology for the development of automotive software systems, and provides the language (meta-model) for their architectural models. It also specifies the architectural modules and functionality of the middleware layer known as the *basic software*. We start by describing the layers of the AUTOSAR reference architecture. We then describe the proposed development methodology by identifying major roles in the automotive development process and the artifacts they produce, with examples of each artifact. We follow up by explaining the role of the AUTOSAR meta-model in the development process and show examples of the architectural models that instantiate this meta-model. We also explain the use of the AUTOSAR meta-model for configuring basic software modules. We conclude the chapter by showing trends in the evolution of the AUTOSAR standard and reflect on its future role in the automotive domain.

4.1 Introduction

The architecture of automotive software systems, as software-intensive systems, can be seen from different views, as presented in Sect. 2.7 and described in more detail by Kruchten in the *4+1 architectural view model* [Kru95]. Two of these architectural views deserve special attention in this chapter, namely the logical and the physical views, so we describe them briefly here as well.

Logical architecture of automotive software systems is responsible for defining and structuring high-level vehicle functionalities such as auto-braking when a pedestrian is detected on the vehicle's trajectory. These functionalities are usually realized by a number of logical software components, e.g., the *PedestrianSensor* component detects a pedestrian and requests full auto-brake from the *BrakeControl* component. These components communicate by exchanging information, e.g., about the pedestrian detected in front of the vehicle. Based on the types of functionalities they realize, logical software components are usually grouped into subsystems that in turn are grouped into logical domains, e.g., active safety and powertrain.

© Springer International Publishing AG 2017
M. Staron, *Automotive Software Architectures*,
DOI 10.1007/978-3-319-58610-6_4

The physical architecture of automotive software systems is usually distributed over a number of computers (today usually more than 100) referred to as Electronic Control Units (ECUs). ECUs are connected via electronic buses of different types (e.g., Can, FlexRay and Ethernet) and are responsible for executing one or several high-level vehicle functionalities defined in the logical architecture. This is done by allocating logical software components responsible for realizing these functionalities to ECUs, thereby transforming them into runnable ECU application software components. Each logical software component is allocated to at least one ECU.

Apart from the physical system architecture that consists of a number of ECUs, each ECU has its own physical architecture that consists of the following main parts:

- Application software that consists of a number of allocated software components and is responsible for executing vehicle functionalities realized by this ECU, e.g., detecting pedestrians on the vehicle's trajectory.
- Middleware software responsible for providing services to the application software, e.g., transmission/reception of data on the electronic buses, and tracking diagnostic errors.
- Hardware that includes a number of drivers responsible for controlling different hardware units, e.g., electronic buses and the CPU of the ECU.

The development of the logical and physical architectural views of automotive software systems and their ECUs is mostly done following the MDA (Model-Driven Architecture) approach [Obj14]. This means that the logical and physical system architecture and the physical ECU architecture are described by means of architectural models. Looking into the automotive architectural design from the process point of view, car manufacturers (OEMs, Original Equipment Manufacturers) are commonly responsible for the logical and physical design of the system, while a hierarchy of suppliers is responsible for the physical design of specific ECUs, implementation of their application and middleware software and the necessary hardware [BKPS07].

In order to facilitate this distributed design and development of automotive software systems and their architectural components, the AUTOSAR (AUTomotive Open Systems ARchitecture) standard was introduced in 2003 as a joint partnership of automotive OEMs and their software and hardware vendors. Today, AUTOSAR consists of more than 150 global partners [AUT16a] and is therefore considered a de facto standard in the automotive domain. AUTOSAR is built upon the following major objectives:

1. Standardization of the reference ECU architecture and its layers. This increases the reusability of application software components in different car projects (within one or multiple OEMs) developed by the same software suppliers.
2. Standardization of the development methodology. This enables collaboration between a number of different parties (OEMs and a hierarchy of suppliers) in the software development process for all ECUs in the system.

3. Standardization of the language (meta-model) for architectural models of the system/ECUs. This enables a smooth exchange of architectural models between different modeling tools used by different parties in the development process.
4. Standardization of ECU middleware (basic software, BSW) architecture and functionality. This allows engineers from OEMs to focus on the design and implementation of high-level vehicle functionalities that can, in contrast to ECU middleware, create competitive advantage.

In the next four sections (4.2–4.6), we show how AUTOSAR achieves each one of these four objectives. In Sect. 4.6, we analyze trends in the evolution of AUTOSAR and how it can be measured. In Sect. 4.7, we present current initiatives regarding the future role of AUTOSAR in the development of automotive software systems. Finally, in the last two sections (4.8 and 4.9), we provide guidelines for further readings and conclude this chapter with a brief summary.

4.2   AUTOSAR Reference Architecture

The architectural design of ECU software based on AUTOSAR is done according to the three-layer architecture that is built upon the ECU hardware layer, as presented in Fig. 4.1.

The first layer, *Application software*, consists of a number of software components that realize a set of vehicle functionalities by exchanging data using interfaces

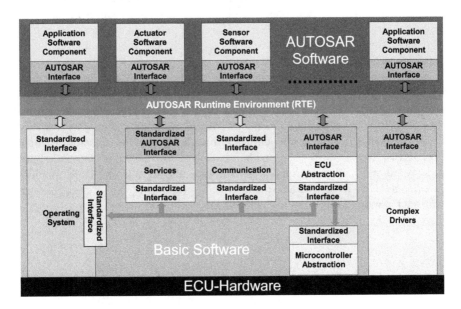

Fig. 4.1 AUTOSAR layered software architecture [AUT16g]

defined on these components (referred to as ports). This layer is based on the logical architectural design of the system. The second layer, *Run-time environment* (RTE), controls the communication between software components, abstracting the fact that they may be allocated onto the same or different ECUs. This layer is usually generated automatically, based on the software component interfaces. If the software components are allocated to different ECUs, transmission of the respective signals on the electronic buses is needed, which is done by the third layer (*Basic software*).

The *Basic software* layer consists of a number of BSW modules and it is responsible for the non-application-related ECU functionalities. One of the most important basic software functionalities is the *Communication* between ECUs, i.e., signal exchange. It consists of BSW modules such as *COM* (Communication Manager) that are responsible for signal transmission and reception. However, AUTOSAR basic software also provides a number of *Services* to the *Application software* layer, e.g., diagnostics realized by *DEM* (Diagnostic Event Manager) and *DCM* (Diagnostic Communication Manager) modules that are responsible for logging error events and transmitting diagnostic messages, respectively, and the *Operating System* for scheduling ECU runnables. The majority of BSW modules are configured automatically, based on the architectural models of the physical system [LH09], e.g., periodic transmission of a set of signals packed into frames on a specific electronic bus.

Communication between higher-level functionalities of ECU *Basic software* and drivers controlling the ECU hardware realized by the *Microcontroler Abstraction* BSW modules is done by the *ECU Abstraction* BSW modules, e.g., bus interface modules such as *CanIf*, which is responsible for the transmission of frames containing signals on the CAN bus. Finally, AUTOSAR provides the possibility for application software components to communicate directly with hardware, thus bypassing the layers of the AUTOSAR software architecture, by means of custom implementations of *Complex Drivers*. This approach is, however, considered non-standardized.

Apart from the *Complex Drivers*, *RTE* and modules of the *Basic Software* layer are completely standardized by AUTOSAR, i.e., AUTOSAR provides detailed functional specifications for each module. This standardization, together with the clear distinction between the *Application software*, *RTE* and *Basic software* layers, allows ECU designers and developers to specifically focus on the realization of high-level vehicle functionalities, i.e., without the need to think about the underlying middleware and hardware. Application software components and BSW modules are often developed by different suppliers who specialize in one of these areas, as explained in more detail in the following section.

4.3   AUTOSAR Development Methodology

On the highest level of abstraction, automotive vendors developing architectural components following the AUTOSAR methodology can be classified into one of the following four major roles in the automotive development process:

- **OEM**: responsible for the logical and physical system design.
- **Tier1**: responsible for the physical ECU design and implementation of the software components allocated to this ECU.
- **Tier2**: responsible for the implementation of ECU basic software.
- **Tier3**: responsible for supplying ECU hardware, hardware drivers and corresponding compilers for building the ECU software.

In most cases, different roles represent different organizations/companies involved in the development process. For example, one car manufacturer plays the role of OEM, two software vendors play the roles of Tier1 and Tier2, respectively, and one "silicon" vendor plays the role of Tier3. However, in some cases, these roles can also be played by the same company, e.g., one car manufacturer plays the role of OEM and Tier1 by doing logical and physical system design, physical ECU design and implementation of the allocated software components (in-house development), or one software vendor plays the role of Tier1 and Tier2 by doing implementation of both the software components and the BSW modules. The development process involving all roles and their tasks is presented in Fig. 4.2.

OEMs start with *logical system design* (1) by modeling a number of composite logical software components and their port interfaces representing data exchange points. These components are usually grouped into subsystems that are in turn

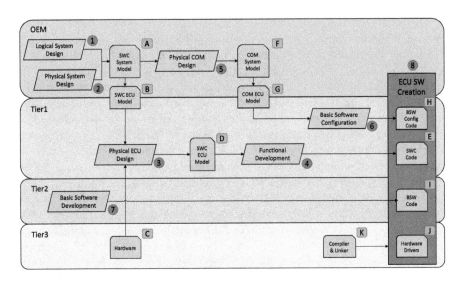

Fig. 4.2 AUTOSAR development process

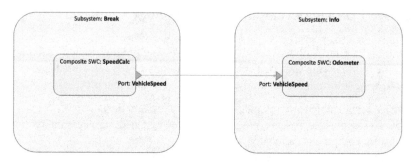

Fig. 4.3 Example of the logical system design done by OEMs (1)

grouped into logical domains. In the later stages of the development process, usually in the *physical ECU design* (3), composite software components are broken down into a number of atomic software components, but this could have been done already in the logical system design phase by OEMs. An example of the logical system design of the minimalistic system created for the purpose of this chapter that calculates vehicle speed and presents its value to the driver is presented in Fig. 4.3.

The example contains two subsystems, *Break* and *Info*, each of which consists of one composite software component, *SpeedCalc* and *Odometer*, respectively. The *SpeedCalc* component is responsible for calculating vehicle speed and it provides this information via the *VehicleSpeed* sender port. The *Odometer* component is responsible for presenting the vehicle speed information to the driver and it requires this information via the *VehicleSpeed* receiver port.

As soon as a certain number of subsystems and software components have been defined in the *logical system design* phase (1), OEMs can start with the *physical system design* (2), which involves modeling a number of ECUs connected using different electronic buses and deployment of software components to these ECUs. In case two communicating software components (with connected ports) are allocated to different ECUs, this phase also involves the creation of system signals that will be transmitted over the electronic bus connecting these two ECUs. An example of the physical system design of our minimalistic system is presented in Fig. 4.4.

The example contains two ECUs, *BreakControl* and *DriverInfo*, connected using the *Can1* bus. The *SpeedCalc* component is deployed to the *BreakControl* ECU while the *Odometer* component is deployed to the *DriverInfo* ECU. As these two components are deployed to different ECUs, information about vehicle speed is exchanged between them in a form of system signal named *VehicleSpeed*.

After the *physical system design* phase (2) is finished, detailed design of the car's functionalities allocated to composite software components deployed to different ECUs (*physical ECU design*) can be performed by the Tier1s (3). As different ECUs are usually developed by different Tier1s, OEMs are responsible for extracting the relevant information about the deployed software components from the generated *SWC system model* (A) into the *SWC ECU model* (B), known as the *ECU Extract*. The main goal of the physical ECU design phase is to break down the composite software components into a number of atomic software components that will in the

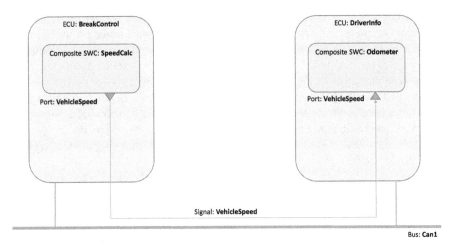

Fig. 4.4 Example of the physical system design done by OEMs (2)

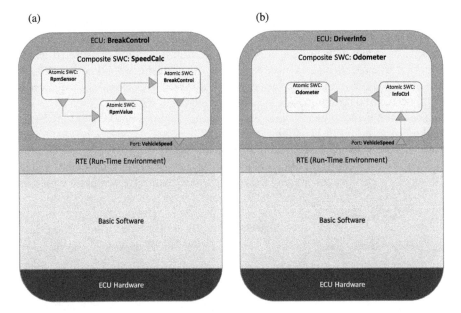

Fig. 4.5 Example of the physical ECU design done by the Tier1s (3). (**a**) BreakControl ECU. (**b**) DriverInfo ECU

end represent runnable entities at ECU run-time. An example of the physical ECU design of our minimalistic system is presented in Fig. 4.5.

The example shows detailing of the *SpeedCalc* and *Odometer* composite software components into a number of atomic software components that will represent runnables in the final ECU software. *SpeedCalc* consists of the *RpmSensor* sensor component that measures the speed of axis rotation, the *RpmValue* atomic software component that calculates the value of the rotation and the *BreakControl* atomic software component that calculates the actual vehicle speed based on the value of the axis rotation. *Odometer* consists of the *InfoControl* atomic software component that receives information about the vehicle speed and the *Odometer* atomic software component that presents the vehicle speed value to the driver.

The ECU design phase is also used to decide upon the concrete implementation data types used in the code for the data exchanged between software components based on the choice of the concrete ECU *hardware* (C) delivered by the Tier3s. For example, data can be stored as floats if the chosen CPU has support for working with the floating points.

Based on the detailed *SWC ECU model* containing the atomic software components (D), the Tier1s can continue with the *functional development* of the car's functionalities (4) allocated onto these components. This is usually done with a help of behavioral modeling with modeling tools such as Matlab Simulink, as explained in Sect. 5.2, that are able to generate source *SWC code* for the atomic software components (E) automatically from the Simulink models [LLZ13]. This part is outside of the AUTOSAR scope.

During the physical ECU design and functional development phases performed by the Tier1s, OEMs can work on the *physical COM design* (5) that aims to complete the system model with packing of signals into frames that are transmitted on the electronic buses. This phase is necessary for configuring the communication (COM) part of the AUTOSAR *basic software configuration* (6). An example of the physical COM design of our minimalistic system is presented in Fig. 4.6.

The example shows one frame of eight bytes named *CanFrm01* that is transmitted by the *BreakControl* ECU on the *Can1* bus and received by the *DriverInfo* ECU. It transports the *VehicleSpeed* signal in its first two bytes.

After the physical COM design phase has been completed for the entire system, OEMs are responsible for creating *COM ECU model* extracts (G) from the generated

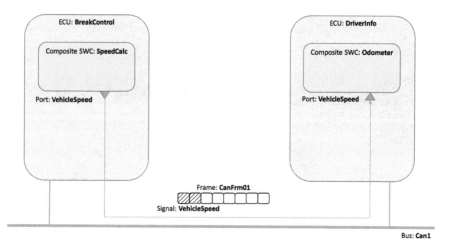

Fig. 4.6 Example of the physical COM design done by OEMs (5)

COM system model (F) for each ECU that contains only ECU-relevant information about the COM design. This step is similar to the step taken after the logical and physical system design, related to the extraction of ECU-relevant information about application software components. These ECU extracts are then sent to the Tier1s, which use them as input for configuring the COM part of the ECU *basic software configuration* (6) and, together with configuring the rest of BSW (diagnostics services, operating system, etc.), generate the complete *BSW configuration code* (H) for the developed ECU. An example of the BSW configuration design of our minimalistic system is presented in Fig. 4.7.

The example shows different groups of BSW modules, i.e., *Operating System*; *Services* including modules such as *DEM* and *DCM*; *Communication* including modules such as *COM*; and *ECU Abstraction* including modules such as *CanIf* needed for the transmission of frames on the Can bus in our example.

The actual ECU *basic software development* (7) is done by the Tier2s, based on the detailed specifications of each BSW module provided by the AUTOSAR standard, e.g., *COM*, *CanIf* or *DEM* modules. The outcome of this phase is a complete *BSW code* (I) for the entire basic software that is usually delivered by the Tier2s in the form of libraries. The *hardware drivers* for the chosen hardware (J), in our example the *CAN* driver, are delivered by the Tier3s.

The last stage in *ECU software creation* (8) is to compile and link the functional *SWC code* (E), the *BSW configuration code* (H), the functional *BSW code* (I) and the *hardware drivers* (J). This is usually done using the *compiler and linker* (K) delivered by the Tier3s.

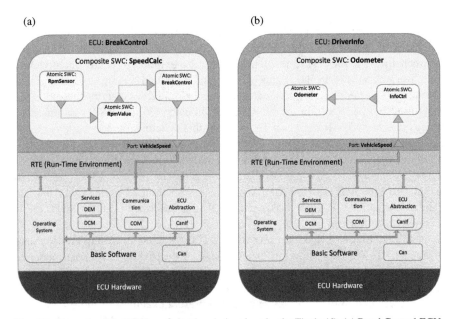

Fig. 4.7 Example of the BSW configuration design done by the Tier1s (6). (**a**) BreakControl ECU. (**b**) DriverInfo ECU

Despite the fact that the described methodology of AUTOSAR is reminiscent of the traditional waterfall development approach, except from the decoupled development of the ECU functional code and the ECU BSW code, in practice it represents just one cycle of the entire development process. In other words, steps (1)–(6) are usually repeated a number of times, adding new functionalities to the system and its ECUs. For example, new composite software components are introduced in the logical system design (1), requiring new signals in the physical system design (2); new atomic software components are introduced as part of the new composite software components in the physical ECU design (3) and implemented in the functional development (4); and new frames to transport the new signals are introduced in the physical COM design (5) and configured in the BSW configuration design (6) phase. Sometimes even the ECU hardware (C) and its compiler/linker (K) and drivers (J) can be changed between different cycles, in case it cannot withstand the additional functionality.

Examples of AUTOSAR based logical system design (1), physical system design (2), physical ECU design (3) and physical COM design (5) are presented in Sect. 4.4. Examples of AUTOSAR-based basic software development (7) and basic software configuration (6) are presented in Sect. 4.5. As already stated, functional development of software components (4) is outside of the scope of AUTOSAR and this chapter.

4.4 AUTOSAR Meta-Model

As we have seen in the previous section, a number of architectural models, as outcomes of different phases in the development methodology, are exchanged between different roles in the development process. In order to ensure that modeling tools used by OEMs in the logical (1), physical (2) and system design communication (5) phases are able to create models that could be read by the modeling tools used by the Tier1s in the physical ECU design (3) and BSW configuration phases (6), AUTOSAR defines a meta-model that specifies the language for these exchanged models [NDWK99]. Therefore, models (A), (B), (D), (F) and (G) represent instances of the AUTOSAR meta-model that specifies their abstract syntax in the UML language. The models themselves are serialized into XML (referred to as ARXML, the AUTOSAR XML), which represents their concrete syntax, and are validated by the AUTOSAR XML schema that is generated from the AUTOSAR meta-model [PB06].

In this section, we first describe the AUTOSAR meta-modeling environment in Sect. 4.4.1. We then show an example use of the AUTOSAR meta-model in the logical system design (1), physical system design (2), physical ECU design (3) and physical COM design (5) phases in Sect. 4.4.2 using our minimalistic system presented in the previous section and present examples of these models in the ARXML syntax. Finally, we discuss the semantics of the AUTOSAR models described in the AUTOSAR template specifications in Sect. 4.4.3.

4.4.1 AUTOSAR Meta-Modeling Environment

As opposed to the commonly accepted meta-modeling hierarchy of MOF [Obj04] that defines four layers [BG01], the AUTOSAR modeling environment has a five-layer hierarchy, as presented below (the names of the layers are taken from the AUTOSAR *Generic Structure* specification [AUT16f]):

1. The **ARM4**: MOF 2.0, e.g., the MOF Class
2. The **ARM3**: UML and AUTOSAR UML profile, e.g., the UML Class
3. The **ARM2**: Meta-model, e.g., the SoftwareComponent
4. The **ARM1**: Models, e.g., the WindShieldWiper
5. The **ARM0**: Objects, e.g., the WindShieldWiper in the ECU memory

The mismatch between the number of layers defined by MOF and AUTOSAR lies in the fact that MOF considers only layers connected by the linguistic instantiation (e.g., *SystemSignal* is an instance of UML *Class*), while AUTOSAR considers both linguistic and ontological layers (e.g., *VehicleSpeed* is an instance of *SystemSignal*) [Küh06]. To link these two interpretations of the meta-modeling hierarchy, we can visualize the AUTOSAR meta-modeling hierarchy using a two-dimensional representation (known as OCA—Orthogonal Classification Architecture [AK03]), as shown in Fig. 4.8. Linguistic instantiation ("L" layers corresponding to MOF layers) are represented vertically and ontological layers ("O" layers) horizontally.

The *ARM2* layer is commonly referred to as the "AUTOSAR meta-model" and it ontologically defines, using UML syntax (i.e., AUTOSAR meta-model is defined as an instance of UML), AUTOSAR models residing on the *M1* layer (both the

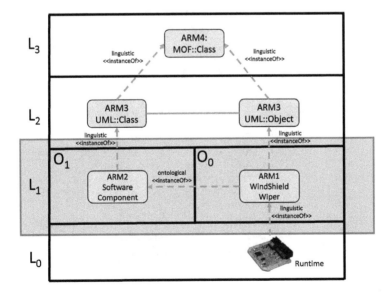

Fig. 4.8 AUTOSAR meta-model layers [DSTH16]

AUTOSAR meta-model and AUTOSAR models are located on the *L1* layer). The AUTOSAR meta-model also uses a UML profile that extends the UML meta-model on the *ARM3* layer, which specifies the used stereotypes and tagged values.

Structurally, the AUTOSAR meta-model is divided into a number of top-level packages referred to as "templates", where each template defines how to model one part of the automotive system. The modeling semantics, referred to as design requirements and constraints, are described in the AUTOSAR template specifications [Gou10]. The AUTOSAR templates and their relations are presented in Fig. 4.9.

Probably the most important templates for the design of automotive software systems are the *SWComponentTemplate*, which defines how to model software components and their interaction; *SystemTemplate*, which defines how to model

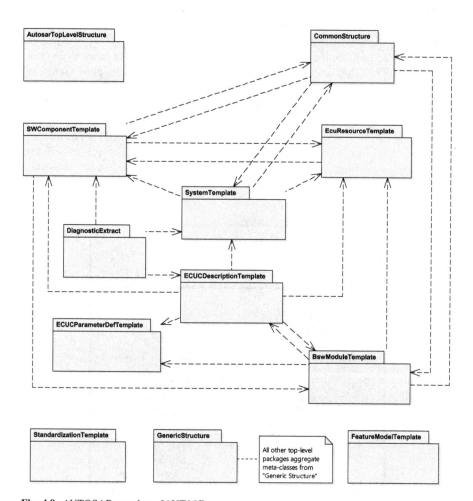

Fig. 4.9 AUTOSAR templates [AUT16f]

ECUs and their communication; and *ECUCParameterDefTemplate* and *ECUCDescriptionTemplate*, which define how to configure ECU basic software. In addition to these templates, AUTOSAR *GenericStructure* template is used to define general concepts (meta-classes) used by all other templates, e.g., handling different variations in architectural models related to different vehicles. In the next subsection, we provide examples of these templates and AUTOSAR models that instantiate them.

4.4.2 Architectural Design Based on the AUTOSAR Meta-Model

A simplified excerpt from the *SWComponentTemplate* that is needed for the logical system and physical ECU design of our minimalistic example that calculates vehicle speed and presents its value to the driver is presented in Fig. 4.10.

The excerpt shows the abstract meta-class *SwComponent* that can be either *AtomicSwComponent* or *CompositeSwComponent*, which may refer to multiple *AtomicSwComponents*. Both types of *SwComponents* may contain a number of *Ports* that can either be *ProvidedPorts* providing data to the other components in the system, or *RequiredPorts* requiring data from the other components in

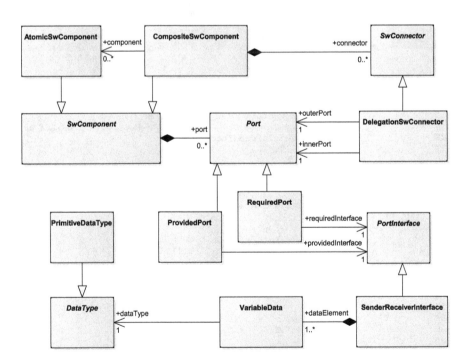

Fig. 4.10 Logical and ECU design example (*SwComponentTemplate*)

the system. Ports on the *CompositeSwComponent*s are connected to the ports of the *AtomicSwComponent*s using *DelegationSwConnector*s that belong to the *CompositeSwComponent*s, i.e., *DelegationSwConnector* points to an *outerPort* of the *CompositeSwComponent* and an *innerPort* of the *AtomicSwComponent*. Finally, *Port*s refer to a corresponding *PortInterface*, e.g., *SenderReceiverInterface* or *ClientServerInterface*, that contains the actual definition of the *DataType* that is provided or required by this port (e.g., unsigned integer of 32 bits or a struct that consists of an integer and a float).

The model of our example of the logical system design presented in Fig. 4.3 that instantiates the *SWComponentTemplate* part of the meta-model is shown in Fig. 4.11 in ARXML syntax. We chose ARXML as it is used as a model exchange format between OEMs and Tier1s, but UML could be used as well.

The example shows the definition of the *SpeedCalc* composite software component (lines 1–11) with the *VehicleSpeed* provided port (lines 4–9), and the *Odometer* composite software component (lines 12–22) with the *VehicleSpeed*

```
1    <COMPOSITE-SW-COMPONENT UUID="...">
2        <SHORT-NAME>SpeedCalc</SHORT-NAME>
3        <PORTS>
4            <PROVIDED-PORT UUID="...">
5                <SHORT-NAME>VehicleSpeed</SHORT-NAME>
6                <PROVIDED-INTERFACE-REF DEST="SENDER-RECEIVER-INTERFACE">
7                    /.../VehicleSpeedInterface
8                </PROVIDED-INTERFACE-REF>
9            </PROVIDED-PORT>
10       </PORTS>
11   </COMPOSITE-SW-COMPONENT>
12   <COMPOSITE-SW-COMPONENT UUID="...">
13       <SHORT-NAME>Odometer</SHORT-NAME>
14       <PORTS>
15           <REQUIRED-PORT UUID="...">
16               <SHORT-NAME>VehicleSpeed</SHORT-NAME>
17               <REQUIRED-INTERFACE-REF DEST="SENDER-RECEIVER-INTERFACE">
18                   /.../VehicleSpeedInterface
19               </REQUIRED-INTERFACE-REF>
20           </REQUIRED-PORT>
21       </PORTS>
22   </COMPOSITE-SW-COMPONENT>
23   <SENDER-RECEIVER-INTERFACE UUID="...">
24       <SHORT-NAME>VehicleSpeedInterface</SHORT-NAME>
25       <DATA-ELEMENTS>
26           <VARIABLE-DATA UUID="...">
27               <SHORT-NAME>VehicleSpeed</SHORT-NAME>
28               <DATA-TYPEREF DEST="PRIMITIVE-DATA-TYPE">
29                   /.../UInt16
30               </DATA-TYPE-REF>
31           </VARIABLE-DATA>
32       </DATA-ELEMENTS>
33   </SENDER-RECEIVER-INTERFACE>
34   <PRIMITIVE-DATA-TYPE UUID="...">
35       <SHORT-NAME>UInt16</SHORT-NAME>
36   </PRIMITIVE-DATA-TYPE>
```

Fig. 4.11 AUTOSAR model example: logical design

required port (15–20). Both ports refer to the same sender-receiver interface (lines 23–33) that in turn refers to the unsigned integer type of 16 bits (lines 34–36) for the provided/required data.

According to the AUTOSAR methodology, these composite software components are, after their allocation to the chosen ECUs, broken down into a number of atomic software components during the physical ECU design phase. The partial model of our minimalistic example of the physical ECU design presented in Fig. 4.5 that instantiates the *SWComponentTemplate* part of the meta-model is shown in Fig. 4.12 in ARXML syntax.

The example shows the definition of the *BreakControl* atomic software component (lines 31–41) with the *VehicleSpeed* provided port (lines 34–39) that is referenced (lines 12–17) from the *SpeedCalc* composite software component (lines 1–30). We can also see the delegation connector *Delegation1* inside the *SpeedCalc*

```
1   <COMPOSITE-SW-COMPONENT UUID="...">
2       <SHORT-NAME>SpeedCalc</SHORT-NAME>
3       <PORTS>
4           <PROVIDED-PORT UUID="...">
5               <SHORT-NAME>VehicleSpeed</SHORT-NAME>
6               <PROVIDED-INTERFACE-REF DEST="SENDER-RECEIVER-INTERFACE">
7                   /.../VehicleSpeedInterface
8               </PROVIDED-INTERFACE-REF>
9           </PROVIDED-PORT>
10      </PORTS>
11      <COMPONENTS>
12          <COMPONENT>
13              <SHORT-NAME>BreakControl</SHORT-NAME>
14              <COMPONENT-REF DEST="ATOMIC-SW-COMPONENT">
15                  /.../BreakControl
16              </COMPONENT-REF>
17          </COMPONENT>
18      </COMPONENTS>
19      <CONNECTORS>
20          <DELEGATION-SW-CONNECTOR UUID="...">
21              <SHORT-NAME>Delegation1</SHORT-NAME>
22              <INNER-PORT-REF DEST="P-PORT-PROTOTYPE">
23                  /.../BreakControl/VehicleSpeed
24              </INNER-PORT-REF>
25              <OUTER-PORT-REF DEST="P-PORT-PROTOTYPE">
26                  /.../SpeedCalc/VehicleSpeed
27              </OUTER-PORT-REF>
28          </DELEGATION-SW-CONNECTOR>
29      </CONNECTORS>
30  </COMPOSITE-SW-COMPONENT>
31  <ATOMIC-SW-COMPONENT UUID="...">
32      <SHORT-NAME>BreakControl</SHORT-NAME>
33      <PORTS>
34          <PROVIDED-PORT UUID="...">
35              <SHORT-NAME>VehicleSpeed</SHORT-NAME>
36              <PROVIDED-INTERFACE-REF DEST="SENDER-RECEIVER-INTERFACE">
37                  /.../VehicleSpeedInterface
38              </PROVIDED-INTERFACE-REF>
39          </PROVIDED-PORT>
40      </PORTS>
41  </ATOMIC-SW-COMPONENT>
```

Fig. 4.12 AUTOSAR model example: ECU design

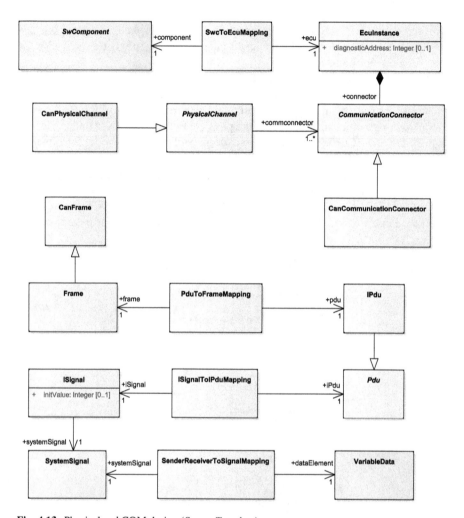

Fig. 4.13 Physical and COM design (*SystemTemplate*)

composite software component (lines 20–28) that connects the provided ports in the *SpeedCalc* and *BreakControl* software components.

A simplified excerpt from the *SystemTemplate* that is needed for the physical and COM system designs of our minimalistic example is presented in Fig. 4.13.

Related to the physical system design, the excerpt shows the *EcuInstance* meta-class with the *diagnosticAddress* attribute which may contain a number of *CommunicationConnectors* that represent connections of *EcuIstance* to a *PhysicalChannel* (e.g., *CanCommunicationConnector* connects one *EcuInstance* to a *CanPhysicalChannel*). A number of *SwComponents* (*CompositeSwComponents* or *AtomicSwComponents*) created in the logical designs can be allocated to one *EcuInstace* by means of *SwcToEcuMappings*.

Related to the physical COM design, the excerpt shows the *SenderReceiver-ToSignalMapping* of the *VariableData* created in the logical design of a *SystemSignal*. It also shows that one *SystemSignal* can be sent to multiple buses by creating different *ISignals* and mapping them to *IPdus*, which are in turn mapped to *Frames*. *IPdu* is one type of *Pdu* (Protocol Data Unit) that is used for transporting signals, and there may be other types of *Pdus*, e.g., *DcmPdu* for transporting diagnostic messages.

The model of our example of the physical system design presented in Fig. 4.4 that instantiates the *SystemTemplate* part of the meta-model is shown in Fig. 4.14.

The example shows the definition of the *BreakControl* ECU with diagnostic address 10 (lines 1–9) that owns a CAN communication connector (lines 5–7). It also shows the mapping of the *SpeedCalc* composite software component onto the *BreakControl* ECU (lines 10–14). Finally, it shows the definition of the *Can1* physical channel (lines 15–24) that points to the CAN communication connector of the *BreakControl* ECU (lines 19–21), thereby indicating that this ECU is connected to *Can1*.

The model of our example of the COM system design presented in Fig. 4.6 that instantiates the *SystemTemplate* part of the meta-model is shown in Fig. 4.15.

The example shows the definition of the *VehicleSpeed* system signal (lines 1–3) that is mapped to the *SpeedCalc* variable data element defined in the logical design phase (lines 4–12). The example also shows the creation of the *ISignal VehicleSpeedCan1* (lines) with initial value of 0 that is meant to transmit the vehicle speed on the *Can1* bus defined in the physical design phase. This *ISignal* is mapped to *Pdu1* (lines 20–22) using *ISignalToIPduMapping* (23–27) that in turn is mapped to *CanFrame1* (lines 28–30) using *IPduToFrameMapping* (lines 31–35).

```
1   <ECU-INSTANCE UUID="...">
2       <SHORT-NAME>BreakControl</SHORT-NAME>
3       <ECU-ADDRESS>10</ECU-ADDRESS>
4       <CONNECTORS>
5           <CAN-COMMUNICATION-CONNECTOR UUID="...">
6               <SHORT-NAME>Can1Connector</SHORT-NAME>
7           </CAN-COMMUNICATION-CONNECTOR>
8       </CONNECTORS>
9   </ECU-INSTANCE>
10  <SWC-TO-ECU-MAPPING UUID="...">
11      <SHORT-NAME>Mapping1</SHORT-NAME>
12      <COMPONENT-REF DEST="SW-COMPONENT">/.../SpeedCalc</SW-REF>
13      <ECU-REF DEST="ECU-INSTANCE">/.../BreakControl</ECU-REF>
14  </SWC-TO-ECU-MAPPING>
15  <CAN-PHYSICAL-CHANNEL UUID="...">
16      <SHORT-NAME>Can1</SHORT-NAME>
17      <COMM-CONNECTORS>
18          <COMMUNICATION-CONNECTOR-REF-CONDITIONAL>
19              <COMMUNICATION-CONNECTOR-REF DEST="CAN-COMMUNICATION-CONNECTOR">
20                  /.../BreakControl/Can1Connector
21              </COMMUNICATION-CONNECTOR-REF>
22          </COMMUNICATION-CONNECTOR-REF-CONDITIONAL>
23      </COMM-CONNECTORS>
24  </CAN-PHYSICAL-CHANNEL>
```

Fig. 4.14 AUTOSAR model example: physical design

```
1   <SYSTEM-SIGNAL UUID="...">
2       <SHORT-NAME>VehicleSpeed</SHORT-NAME>
3   </SYSTEM-SYGNAL>
4   <SENDER-RECEIVER-TO-SIGNAL-MAPPING UUID="...">
5       <SHORT-NAME>Mapping2</SHORT-NAME>
6       <DATA-ELEMENT-REF DEST="VARIABLE-DATA">
7           /.../VehicleSpeedInterface/SpeedCalc
8       </DATA-ELEMENT-REF>
9       <SYSTEM-SIGNAL-REF DEST="SYSTEM-SIGNAL">
10          /.../VehicleSpeed
11      </SYSTEM-SIGNAL-REF>
12  </SENDER-RECEIVER-TO-SIGNAL-MAPPING>
13  <I-SYGNAL UUID="...">
14      <SHORT-NAME>VehicleSpeedCan1</SHORT-NAME>
15      <INIT-VALUE>0</INIT-VALUE>
16      <SYSTEM-SIGNAL-REF DEST="SYSTEM-SIGNAL">
17          /.../VehicleSpeed
18      </SYSTEM-SIGNAL-REF>
19  </I-SYGNAL>
20  <I-PDU UUID="...">
21      <SHORT-NAME>IPdu1</SHORT-NAME>
22  </I-PDU>
23  <I-SIGNAL-TO-I-PDU-MAPPING UUID="...">
24      <SHORT-NAME>Mapping3</SHORT-NAME>
25      <I-PDU-REF DEST="I-PDU">/.../IPdu1</I-PDU-REF>
26      <I-SIGNAL-REF DEST="I-SIGNAL">/.../VehicleSpeedCan1</I-SIGNAL-REF>
27  </I-SIGNAL-TO-I-PDU-MAPPING>
28  <CAN-FRAME UUID="...">
29      <SHORT-NAME>CanFrame1</SHORT-NAME>
30  </CAN-FRAME>
31  <I-PDU-TO-FRAME-MAPPING UUID="...">
32      <SHORT-NAME>Mapping4</SHORT-NAME>
33      <PDU-REF DEST="I-PDU">/.../IPdu1</PDU-REF>
34      <FRAME-REF DEST="CAN-FRAME">/.../CanFrame1</FRAME-REF>
35  </I-PDU-TO-FRAME-MAPPING>
```

Fig. 4.15 AUTOSAR model example: COM design

4.4.3 AUTOSAR Template Specifications

Like other language definitions, the AUTOSAR meta-model defines only syntax for different types of architectural models, without explaining how its meta-classes shall be used to achieve certain semantics. This is done in natural language specifications called templates [Gou10], e.g., *SwComponentTemplate* and *SystemTemplate*, that explain different parts of the AUTOSAR meta-model. These templates consist of the following main items:

- Design requirements that should be fulfilled by the models (specification items).
- Constraints that should be fulfilled by the models and checked by modeling tools.
- Figures explaining the use of a group of meta-classes.
- Class tables explaining meta-classes and their attributes/connectors.

As an example of a specification item related to our minimalistic example that calculates vehicle speed and presents its value to the driver, we present specification item no. 01009 from the *SystemTemplate* that describes the use of *CommunicationConnector*s:

[TPS_SYST_01009] Definition of CommunicationConnector [An EcuInstance uses CommunicationConnector elements in order to describe its bus interfaces and to specify the sending/receiving behavior.]

As an example of a constraint, we present constraint no. 1032 from the *SwComponentTemplate* that describes the limitation in the use of *DelegationSwConnector*s.

[constr_1032] DelegationSwConnector can only connect ports of the same kind [A DelegationSwConnector can only connect ports of the same kind, i.e. ProvidedPort to ProvidedPort and RequiredPort to RequiredPort.]

The majority of constraints including *constr_1032* could be specified directly in the AUTOSAR meta-model using OCL (Object Constraint Language). However, due to the complexity of OCL and thousands of automotive engineers in more than hundred OEM and supplier companies that develop automotive software components based on AUTOSAR, natural language specifications are considered a better approach for such a wide audience [NDWK99].

Meta-model figures show relationships between a number of meta-classes using UML notation and they are similar to Figs. 4.10 and 4.13 presented in the previous section. These figures are usually followed by class tables that describe the meta-classes in the figures in more detail, e.g. description of the meta-classes, their parent classes and attributes/connectors, so that the readers of the AUTOSAR specification do not need to look directly into the AUTOSAR meta-model which is maintained by the *Enterprise Architect* tool.

In addition to specification items, constraints, figures and class tables, AUTOSAR template specifications also contain a substantial amount of plain text that provides additional explanations, e.g., introductions to the topic and notes about specification items and constraints.

4.5 AUTOSAR ECU Middleware

AUTOSAR provides detailed functional specifications for the modules of its middleware layer (basic software modules). For example, the *COM* specification describes the functionality of the Communication Manager module that is mostly responsible for handling the communication between ECUs, i.e., transmitting signals received from the RTE onto electronic buses and vice versa. These specifications consist of the following main items:

- Functional requirements that should be fulfilled by the implementation of the BSW modules.
- Description of APIs of the BSW modules.

- Sequence diagrams explaining the interaction between BSW modules.
- Configuration parameters that are used for configuring the BSW modules.

The functional side of the AUTOSAR BSW module specifications (functional requirements, APIs and sequence diagrams) is outside of the scope of this chapter. However, we do describe here the general approach to configuration of the BSW modules as it is done based on the AUTOSAR meta-model and its templates.

Two of the AUTOSAR templates are responsible for specifying configuration of the AUTOSAR basic software—*ECUCParameterDefTemplate* and *ECUCDescriptionTemplate* on the *ARM2* layer. *ECUCParameterDefTemplate* specifies the general definition of configuration parameters, e.g., that parameters can be grouped into containers of parameters and that they can be configured at different configuration times (e.g., before or after building the complete ECU software). *ECUCDescriptionTemplate* specifies modeling of concrete parameter and container values that reference their corresponding definitions from the *ECUCParameterDefTemplate*.

The values of configuration parameters from the *ECUCDescriptionTemplate* models can be automatically derived from the models of other templates, e.g., *SoftwareComponentTemplate* and *SystemTemplate*. This process is called "upstream mapping" and it can be done automatically with support from the ECU configuration tools [LH09]. A simplified example of the *ECUCParameterDefTemplate* and *ECUCParameterDefTemplate* and their models, including the upstream mapping process, is shown in Fig. 4.16 in UML syntax.

The *ECUCParameterDefTemplate* on the *ARM2* layer (left blue box) specifies modeling of the definition of configuration parameters (*ECUCParameterDefs*) and containers (*ECUCContainerDefs*), with an example of the integer parameter definition (*ECUCIntegerParameterDef*). The *ECUCDescriptionTemplate* (left yellow box) specifies modeling of the values of containers (*ECUCContainerValues*) and

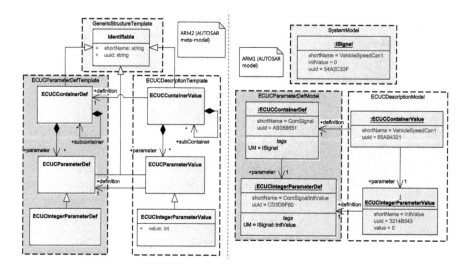

Fig. 4.16 Example of the AUTOSAR templates and their models

parameters (*ECUCParameterValue*s), with an example of the integer parameter value (*ECUCIntegerParameterValue*). As with the elements from the *SwComponentTemplate* and the *SystemTemplate*, the elements from these two templates are also inherited from the common element in the *GenericStructureTemplate* (green box) named *Identifiable*, which provides them with a short name and unique identifier (UUID).

The standardized model (i.e., provided by AUTOSAR) of the *ECUCParameterDefTemplate* can be seen on the *ARM1* layer (right blue box). It shows the *ECUCContainerDef* instance with *shortName* "ComSignal" that refers to the *ECUCParameterDef* instance with *shortName* "ComSignalInitValue". These two elements both have the tagged value named *UM*, denoting Upstream Mapping. The *UM* tagged value for the "ComSignal" container instance refers to the *ISignal* meta-class from the *SystemTemplate*. The *UM* tagged value for the "ComSignalInitValue" parameter instance refers to the *initValue* attribute of the *ISignal*. This implies that for every *ISignal* instance in the *SystemModel*, one *ECUCContainerValue* instance in the *ECUCDescriptionModel* shall be created with an *ECUCParameterValue* instance. The value of this parameter instance shall be equal to the *initValue* attribute of that *SystemSignal* instance.

Considering the "VehicleSpeedCan1" *ISignal* with "initValue" 0 (orange box) that we defined in our *SystemModel* shown in Fig. 4.15 (COM design phase), the *ECUCDescriptionModel* (right yellow box) can be generated. This model contains one instance of the *ECUCContainerValue* with *shortName* "VehicleSpeedCan1" that is defined by the "ComSignal" container definition and refers to one instance of the *ECUCParameterValue* with *shortName* "initValue" of value 0 that is defined by the "ComSignalInitValue" parameter definition.

AUTOSAR provides the standardized *ARM1* models of the *ECUCParameterDefTemplate* for all configuration parameters and containers of the ECU basic software. For example, the *ComSignal* container with *ComSignalInitValue* are standardized for the *COM* BSW module. On the smallest granularity, standardized models of the *ECUCParameterDefTemplate* are divided into a number of packages, where each package contains configuration parameters of one *Basic software* module. At the highest level of granularity, these models are divided into different logical packages, including ECU communication, diagnostics, memory access and IO access.

4.6 AUTOSAR Evolution

The development of the AUTOSAR standard started in 2003 and its first release in mass vehicle production was R3.0.1 from 2007. In this section, we show trends in the evolution of the AUTOSAR meta-model and its requirements (both template and basic software) from its first release until release 4.2.2 from 2016, with a focus on the newer releases (R4.0.1–R4.2.2).

4.6.1   AUTOSAR Meta-Model Evolution

From the architectural modeling point of view, the most important artifact to be analyzed through different releases of AUTOSAR is the AUTOSAR meta-model, as it defines several different types of AUTOSAR models such as SWC, COM and BSW configuration models. We start the analysis of the AUTOSAR meta-model evolution by showing its increase in size from the initial release until the latest one. Figure 4.17 shows the number of AUTOSAR meta-classes in all of its templates (left) and the number of standardized BSW configuration parameters, as instances of the *ECUCParameterDefTemplate* (right) [DSTH14].

The figure indicates relatively even evolution of the AUTOSAR application software and AUTOSAR basic software, except for the R1.0, where no BSW configuration parameters have been standardized. We can also ascertain that there is a significant increase in the number of meta-classes and configuration parameters starting from R4.0.1, and relatively small change between R3.0.1 and R3.1.5. This is because the development of AUTOSAR R4.0.1 started in parallel to that of R3.1.1 and continued to develop as independent branches, namely the 3.x branch and the 4.x branch, for a couple of years until all AUTOSAR OEMs switched to 4.x. The main development focus, such as the introduction of new AUTOSAR features, was on the 4.x branch, and the 3.x branch was considered to be in the maintenance mode, focusing mostly on fixing errors in the meta-model and specifications and implementing the most important features from the 4.x branch.

AUTOSAR R4.0.1 was made public in 2009 and brought significant changes to almost all specifications, including the AUTOSAR meta-model. These changes included a number of new features, referred to as concepts by AUTOSAR, such as for supporting the LIN 2.1 electronic bus and concept enabling the existence of different variants in AUTOASAR models related to different vehicle lines. However, it also included clean-up activities of the meta-classes and configuration parameters of unused or broken concepts. Because AUTOSAR 4.x releases are used today by the majority of AUTOSAR OEMs, we provide in the rest of this section a more detailed analysis of the evolution trends of the 4.x branch. Figure 4.18 shows the number of added, modified and removed meta-classes between different releases of 4.x.

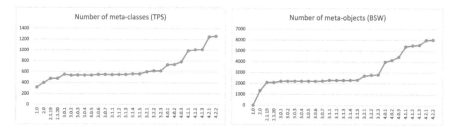

Fig. 4.17 Number of classes (TPS) and objects (BSW)

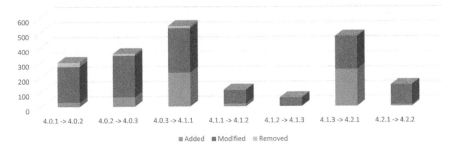

Fig. 4.18 Number of added, modified and removed classes (TPS)

At least three important conclusions about the evolution of the AUTOSAR meta-model can be derived from this figure. First, we can see that the evolution is mostly driven by modifications and additions of meta-classes, while removals are very seldom. The main reason behind this is the strong requirement for backwards compatibility of the AUTOSAR schema that is generated from the AUTOSAR meta-model, e.g., R4.0.2 models should be validated by the R4.0.3 schema or later. Second, we can see that the initial three releases, R4.0.1 to R4.1.1, all have increased the number of added, modified and removed meta-classes, indicating that it took a couple of releases to stabilize the 4.x branch. Finally, we can see that generally only minor AUTOSR releases (second digit changed, e.g., R4.1.1 vs. R4.2.1) bring a lot of new meta-classes. This is related to the AUTOSAR's policy that only major (first digit changed) and minor (second digit changed) releases may introduce new features while revisions (third digit changed) are mostly responsible for fixing errors in the meta-model related to the existing features.

In order for the automotive engineers to be able to make a preliminary assessment of the impact of adopting a new AUTOSAR release, or a subset of its new features, on the used modeling tools, a measure of meta-model change (*NoC*—Number of Changes) has been developed [DSTH14]. *NoC* considers all possible changes to the meta-classes, meta-attributes, and meta-connectors that need to be implemented by the vendors of the AUTOSAR tools used by OEMs and Tier1s. We use this measure to present the estimated effort needed to update the AUTOSAR modeling tool-chain in order to switch from one AUTOSAR release to another (e.g., 3.x to 4.x). The measurement results are presented in Fig. 4.19.

As expected, the highest effort is needed when switching from the older AUTOSAR releases (e.g., R1.0 and R2.0) to the newer ones (e.g., R4.2.1 and R4.2.2). However, the figure also shows that there is a significantly higher effort caused by the AUTOSAR releases in branch 4.x than in releases from the previous branches. This indicates that the functional "big bang" of AUTOSAR started with branch 4.x and it still continues to expand.

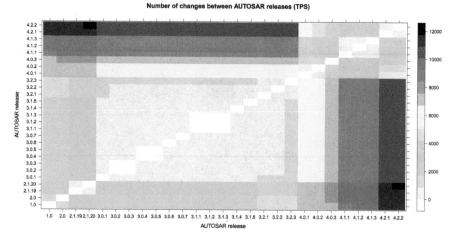

Fig. 4.19 Number of changes between releases (TPS)

Due to the fact that this expansion is mostly driven by new features incorporated into the minor releases of the AUTOSAR branch 4.x, we continue our analysis of the AUTOSAR evolution by presenting the impact of 14 new concepts of AUTOSAR R4.2.1. A brief description of each concept is presented below (more details can be found in [AUT16i]).

1. **SwitchConfiguration**: Enables full utilization of Ethernet and Ethernet switches as a communication medium between ECUs.
2. **SenderReceiverSerialization**: Enables mapping of complex data to single signal entities by means of byte array serialization. The goal is to reduce the number of signals and minimize the signal processing time.
3. **CANFD**: Introduces a new communication protocol for the CAN bus with higher bandwidth and payload for large signals.
4. **EfficientCOMforLargeData**: Enables faster transmission of large signals through the ECU by avoiding overhead of the *COM* module.
5. **E2E Extension**: Reworks the modeling of safety communication (e.g., indicator about modified or missing data during transport) between ECUs so that it does not require additional non-standardized code.
6. **GlobalTimeSynchronization**: Provides a common time base that is distributed across various buses for accurate ECU data correlation.
7. **SupportForPBLAndPBSECUConfiguration**: Enables simultaneous configuration of several ECU variants in a car and different car lines.
8. **SecureOnboardCommunication**: Provides mechanisms for securing the communication on in-vehicle networks (e.g., communication between the car and the outside world).
9. **SafetyExtensions**: Provides mechanisms to realizing and documenting functional safety of AUTOSAR systems (e.g. according to ISO 26262).

10. **DecentralizedConfiguration**: Extension of the AUTOSAR meta-model to support transfer of diagnostic needs of OEMs to suppliers by use of AUTOSAR-complaint models.
11. **IntegrationOfNonARSystems**: Enables integration of non-AUTOSAR systems, e.g., Genivi, into an AUTOSAR system during development.
12. **NVDataHandlingRTE**: Provides efficient mechanisms for the software components to handle non-volatile data.
13. **EcuMFixedMC**: Provides support for ECU state handling on ECUs with multiple cores.
14. **AsilQmProtection**: Provides means for protecting modules developed according to safety regulations from other potentially unsafe modules (i.e., it reduces the chance of error propagation to safety-critical modules).

Figure 4.20 shows the results of the *NoC* measure calculated for each of these 14 features for all AUTOSAR meta-model templates.

This figure shows two important aspects of the AUTOSAR meta-model evolution related to new concepts. First, they have very different impact on the AUTOSAR meta-model (see left part of the figure), i.e., some concepts have no impact at all, such as the concept of *IntegrationOfNonARSystems*, while some have a significant impact, such as the concept of *DecentralizedConfiguration*. Second, we can see that the vast majority of changes in R4.2.1 is related to new concepts (see right part of the figure) and only a small part to other changes, e.g., fixes of errors related to existing concepts.

Finally, in order to present the results of role-based assessment of impact of the AUTOSAR R4.2.1 concepts, we considered the following design roles in AUTOSAR-based development, whose work is described in Sect. 4.3 [DST15]:

- *Application Software Designer:* Responsible for the definition of software components and their data exchange points (involved in the phases of physical system and physical ECU design; see (2) and (3) in Fig. 4.2).

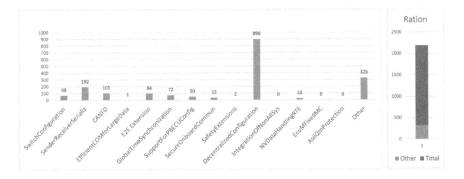

Fig. 4.20 Number of changes per concept (TPS)

- *Diagnostic Designer:* Responsible for the definition of diagnostic services required by software components (involved in the phases of physical system and physical ECU design; see (2) and (3) in Fig. 4.2).
- *ECU Communication Designer:* Responsible for the definition of signals and their transmission on electronic buses in different frames (involved in the phase of physical COM design; see (5) in Fig. 4.2).
- *Basic Software Designer:* Responsible for the design of BSW modules and their interfaces (involved in the phase of BSW development; see (7) in Fig. 4.2).
- *COM Configurator:* Responsible for the configuration of communication BSW modules (involved in the phase of BSW configuration; see (6) in Fig. 4.2).
- *Diagnostic configurator:* Responsible for the configuration of diagnostic BSW modules (involved in the phase of BSW configuration; see (6) in Fig. 4.2).

Figure 4.21 shows the impact of 13 new concepts of AUTOSAR R4.2.1 (all except the concept of *SupportForPBLAndPBSECUConfiguration,* as it causes a significantly higher number of changes in comparison to other concepts, thereby obscuring the results) on these six roles.

This figure shows the following interesting points. First, the *COM Configurator* role followed by the *ECU Communication Designer* role are mostly affected by the concepts, i.e., roles related to the communication between different ECUs. Then, we can see that the majority of concepts do not have impact on all roles, except for the concept of *DecentralizedConfiguration.* Finally, we can see that some concepts, e.g., *IntegrationOfNonARSystems* and *AsilQmProtection,* do not have impact on any of the major roles. The concept of *IntegrationOfNonARSystems* represents a methodological guideline without the actual impact on the models while the concept of *AsilQmProtection* has impact on other safety-related basic software modules that are not explicitly related to ECU communication and diagnostics.

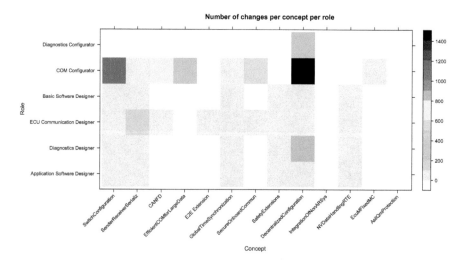

Fig. 4.21 Number of changes per concept affecting different roles

4.6.2 AUTOSAR Requirements Evolution

As already explained, AUTOSAR defines two main types of requirements:

- Design requirements in the AUTOSAR template specifications (*TPS require-ments*) that define semantics for the AUTOSAR meta-model elements. They include both specification items and constraints checked by the modeling tools.
- Functional middleware requirements in the AUTOSAR basic software specifi-cations (*BSW requirements*) that define functionality of the AUTOSAR BSW modules, e.g., *Com, Dem* and *Dcm*.

The evolution of the AUTOSAR TPS requirements is tightly related to the evolution of the AUTOSAR meta-model, as the introduction, modification and removal of meta-classes also require introduction, modification and removal of the supporting requirements for their use. This evolution mostly affects the work of OEMs and Tier1s in the system, ECU and COM design phases of the development. The evolution of the AUTOSAR BSW requirements indicates the functional changes in the ECU middleware that is developed by Tier2s in the basic software development phase and to some extent Tier3s related to the development of drivers for the chosen hardware. In this subsection, we present the analysis of evolution of both types of AUTOSAR requirements for different AUTOSAR releases in branch 4.x [MDS16]. We first show in Fig. 4.22 the increase in the number of requirements in the AUTOSAR templates (left) and BSW specifications (right).

There are two interesting points that can be observed from this figure. First, we can see a constant increase in the number of both TPS and BSW requirements in the new AUTOSAR releases. This indicates that the standard is still growing, i.e., new features are being incorporated into the standard. Second, we can see a relatively steady increase in the number of BSW requirements without disruptive changes in their number between consecutive releases. On the other hand, we can see a big increase (almost three times) in the number of TPS requirements between AUTOSAR R4.0.2 and R4.0.3 and then again almost a double increase between R4.0.3 and R4.1.1 that indicates that the evolution of the TPS and BSW requirements does not follow the same trend. This large increase in the

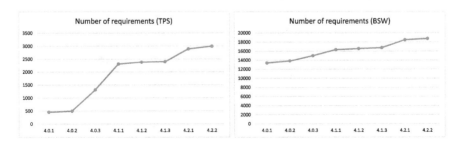

Fig. 4.22 Number of TPS and BSW requirements

number of TPS requirements is partially related to the immaturity of the older
AUTOSAR template specifications, where a lot of plain text had to be converted
into specification items and constraints.

We already showed that AUTOSAR continues to grow by standardizing new
features. In order to assess the influence of new features on the evolution of both
template and BSW specifications, we analyzed the number of added, modified and
removed TPS and BSW requirements. Figure 4.23 shows the results for the TPS
requirements for different releases of the AUTOSAR 4.x branch.

We can see that the evolution of the AUTOSAR TPS specifications is mostly
driven by introduction of new requirements (specification items and constraints).
This is especially the case with minor releases of AUTOSAR, but also in R4.0.3.
Removal of TPS requirements is not common, similarly to removal of meta-classes
from the AUTOSAR meta-model, as it may affect backwards compatibility of the
AUTOSOAR models that is kept high within one major release.

Figure 4.24 shows the number of added, modified and removed BSW requirements for different releases of the AUTOSAR 4.x branch.

The results confirm that the evolution of the AUTOSAR BSW specifications is
also mostly driven by the introduction of new BSW requirements (i.e., new basic
software features). However, we can see that there are generally more modifications

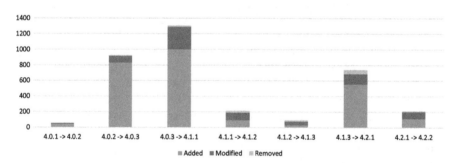

Fig. 4.23 Number of added, modified and removed requirements (TPS)

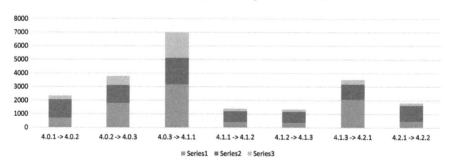

Fig. 4.24 Number of added, modified and removed requirements (BSW)

and removals of BSW requirements than was the case with TPS requirements. High modification to the BSW requirements could indicate lower stability of the AUTOSAR basic software. High removals of the BSW requirements could indicate that certain features become obsolete, probably due to the introduction of newer features that provide the same or similar functionality. Another reason behind more removals of BSW requirements in comparison to TPS requirements is related to the relaxed backwards-compatibility requirements of the AUTOSAR basic software in comparison to the AUTOSAR models, as AUTOSAR models are exchanged between several roles in the development process (e.g., OEM and Tier1) while basic software modules are usually developed by the role of Tier2 only.

4.7 Future of AUTOSAR

The results of the AUTOSAR evolution analysis presented in the previous section show that it is strongly driven by innovations in the form of new features incorporated into the standard. This trend is expected to continue in the future and will be expanded, considering the plans of AUTOSAR for maintaining two platforms in parallel—a classic platform that represents the continuation of the work on the branch 4.x and an adaptive platform that represents a new platform with the goal to satisfy future needs of the automotive industry. The classic platform aims to stabilize and improve the existing AUTOSAR features while the adaptive platform aims to anticipate the future by identifying technological trends and key features for AUTOSAR.

The subsequent release of the classic platform was R4.3.0, planned for the last quarter of 2016. Examples of new concepts that will be supported in this release are presented below:

1. *MacroEncapsulationOfLibraryCalls:* Simplifies handling of interpolation routines provided by libraries (automated selection and parametrization).
2. *CryptoInterface:* Develops the strategy for security SW-/HW-Interface to support technology trends such as Car-2-X communication and autonomous drive.
3. *V2XSupport:* Enables implementation of Intelligent Transportation Systems (ITS) applications [ETS16] as AUTOSAR software components and their integration into an AUTOSAR ECU (Ethernet-based V2X-stack).
4. *ProfilesForDataExchangePoints:* Improves the interoperability between AUTOSAR tools by providing means for describing which data is expected for a given data exchange point.
5. *DecentralizedConfigurationExt01:* Extends the concept of "Decentralized Configuration", which provides a top-down configuration of diagnostics via diagnostic extract, with features such as On-Board Diagnostics (OBD).
6. *ExtendedBufferAccess:* Extends the existing rapid prototyping functionality (quick validation of a software algorithm in an ECU context without the need for a production build) with support for bypassing the RTE.

7. *PolicyManager:* Allows specification of security policies in AUTOSAR ARXML (e.g., insurance company can read data but not modify it).
8. *DLTRework:* Improves the Development Error Tracer (DLT) module that provides generic logging and tracing functionality for the software components, RTE, DEM module, etc.
9. *SOMEIPTransportProtocol:* Defines a protocol for segmentation of Scalable Service-Oriented Middleware over IP (SOME/IP) [Völ13] packets that are larger than 128 kBytes.

Together with these new features, the release will bring improvements to the existing features by fixing a number of issues in the related specifications.

The first release of the adaptive platform is planned for the first quarter of 2017 and will be following a different naming schema that consist of the release year and month, e.g., *R17-03*. Its main goal is to ensure the fulfillment of the 2020 expectations from the automotive industry which state that all major vehicle innovations will be driven by electrical systems. The list of selected main functional drivers for the AUTOSAR's adaptive platform is presented below [AUT16b]:

1. *Highly automated driving:* Support driving automation levels 3–4 according to the NHSTA (National Highway Safety Traffic Administration) [AUT13], i.e., limited driving automation where the driver is occasionally expected to take the control and full driving automation where the vehicle is responsible for performing the entire trip. This includes support for cross-domain computing platforms, high-performance micro-controllers, distributed and remote diagnostics, etc. The levels of autonomous functionality are described further in Sect. 9.2.
2. *Car-2-X applications:* Support interaction of vehicles with other vehicles and off-board systems. This includes support for designing automotive systems with non-AUTOSAR ECUs based on Genivi, Android, etc.
3. *Vehicle in the cloud:* Support vehicle to cloud communication. This includes the development of secured on-board communication, security architecture and secure cloud interaction.
4. *Increased connectivity:* Support increased connectivity of the automotive software systems and other non-AUTOSAR and off-board systems. This includes support for dynamic deployment of software components and common methodology of work regardless of whether the system is on-board or off-board.

The idea behind adaptive cars is depicted in Fig. 4.25 [AUT16b]. The figure shows several classic AUTOSAR ECUs ("C") that are responsible for common vehicle functionalities, e.g., engine or brake control units. The figure also shows several non-AUTOSAR ECUs ("N") that are responsible for infotainment functionalities or communication with the outside world (e.g., Genivi or Android ECUs). Finally, the figure shows certain adaptive AUTOSAR ECUs ("A") that are responsible for realization of advanced car functionalities that usually require inputs or provide outputs to both classic and non-AUTOSAR ECUs, such as car-2-X applications. These ECUs are commonly developed following agile development methodologies and require more frequent updates and run-time configuration.

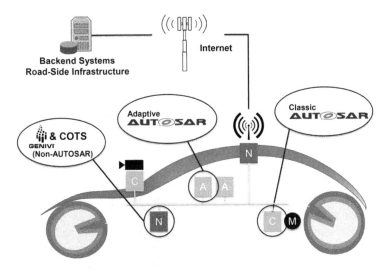

Fig. 4.25 Adaptive AUTOSAR vehicle architecture

Considering the functional drivers for the adaptive platform and the idea behind the adaptive vehicle architecture explained above, adaptive ECUs are expected to be designed using the following principles and technologies (the list is not exhaustive):

- Agile software development methodology based on dynamic software updates. This enables a continuous functional development mode that starts with a minimum viable product.
- Fast addition of new features (application software components) deployed in different packages. This enables fast software innovation cycles.
- Secured service-oriented point-to-point communication. This enables dynamic updates of application software, where new software components subscribe to existing services via a service discovery protocol.
- Wireless updates of the application software. This enables "on the road" software updates without the need for taking the car to a workshop.
- Support for run-time configuration. This enables dynamic adaption of the system based on available functionality.
- High bandwidth for inter-ECU communication (Ethernet). This enables faster transmission of large data.
- Switched networks (Ethernet switches). This enables smart data exchange between different Ethernet buses.
- Micro-processors with external memory instead of micro-controllers. This enables higher amounts of memory and peripherals that can be extended.
- Multi-core processors, parallel computing and hardware acceleration. This enables faster execution of vehicle functions.

- Integration with classic AUTOSAR ECUs or other non-AUTOSAR ECUs (e.g., Genivi, Android). This enables unanimous design of heterogeneous automotive software systems.
- Execution models of access freedom, e.g. full access or sandboxing. This enables security mechanism for separating running programs from each other, e.g., safety- and security-critical programs from the rest.

AUTOSAR plans to achieve this using the adaptive ECU architecture presented in Fig. 4.26 [AUT16b]:

As was the case with classic AUTOSAR platform, AUTOSAR standardizes the middleware layer of the adaptive platform that is referred to as the *Adaptive AUTOSAR Services* layer. However, this layer is organized in functional clusters rather than as a detailed description of the modules (internal structure of the clusters), which enables platform-independent design of the software architectures. Orange clusters represent parts of the ECU architecture that will be standardized in the first release of the adaptive platform, while gray clusters represent parts of the ECU architecture that will be standardized in the later releases.

We can see that, apart from the *Operating system* that is based on POSIX and standardized *Bootloader* for downloading software to the ECU, AUTOSAR plans to deliver *Execution Management*, *Logging and Tracing*, *Diagnostics* and *Communications* functional clusters. The *Execution Management* cluster is responsible for starting/stopping applications related to different car modes and it is based on threads rather than runnables. The *Logging and Tracing* cluster is responsible for collecting information about different events such as the ones related to safety or

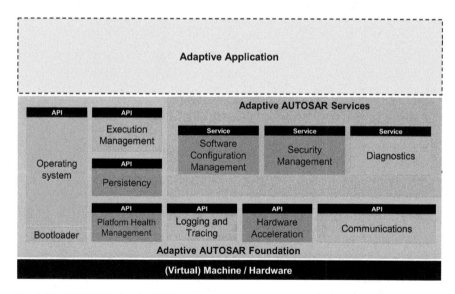

Fig. 4.26 AUTOSAR adaptive ECU architecture

security. As in the classic AUTOSAR platform, the *Diagnostics* cluster is responsible for collection of diagnostic event data which is now possible to exchange with the diagnostic backend. Finally, the *Communications* cluster is responsible for service-oriented communication between ECUs connected via Ethernet (SOME/IP protocol).

The big improvement in the standardization process of the AUTOSAR adaptive platform is the validation of new features before their standardization. This means that AUTOSAR will form a group of engineers that will create prototypes of the new features based on specifications that are planned to be released and provide feedback to AUTOSAR about their feasibility. This ensures agility of the development process within the AUTOSAR consortium as well.

4.8   Further Reading

For those who would like details about AUTOSAR, it is important to understand that AUTOSAR is a huge standard with over 200 specifications and more than 20,000 requirements, so it is nearly impossible to be an expert in all of its features. AUTOSAR's specifications are divided into *standard* and *auxiliary* specifications, where only the standardized ones are required to be followed for achieving full AUTOSAR compliance. Nevertheless, both standardized and auxiliary specifications could be of interest to the readers who would like to learn specifics about the AUTOSAR standard.

We recommend all AUTOSAR beginners to start reading the *Layered Software Architecture* document [AUT16g], as it defines high-level features of AUTOSAR that should be known before diving deeper into other specifications. The AUTOSAR's *Methodology* specification [AUT16h] could be a natural continuation as it contains descriptions of the most important artifacts that are created by different roles in the AUTOSAR development process. However, it also contains many details that may not be understandable at this point, so it should be skimmed through, with us focusing on the familiar topics.

The rest of the readings are specific to the interest topic of the reader. Readers interested in the architectural design of automotive software systems should look into AUTOSAR's template specifications (TPSs). For example, if they are interested in the logical system/ECU design, they should take a look at the AUTOSAR *Software Component* template [AUT16j] in order to understand how to define application software components and their data exchange points. Some general concept used in all templates could be found in the *Generic Structure* template [AUT16f], but it is probably best to follow references from the template that is being read to the concrete section in the *Generic Structure* templates. This is because understanding the entire document at once could be challenging. There is no real need to look at the actual AUTOSAR meta-model specified in UML, as all relevant information and diagrams are exported to the AUTOSAR template specifications.

Readers interested in the functionalities of the AUTOSAR basic software should read the software specifications (SWS) of the relevant basic software modules. For example, if they are interested in the ECU diagnostic functionality, they should take a look at the AUTOSAR *Diagnostic Event Manager* [AUT16d] and *Diagnostic Configuration Manager* [AUT16c] specifications. Requirements applicable to all basic software modules can be found in the *General Requirements on Basic Software Modules* specification [AUT16e].

On a higher granularity level, design requirements from the TPS specifications can be traced to the more formalized requirements from the requirements specifications (RS) documents. Similarly, functional basic software requirements from the SWS specifications can be traced to the more formalized requirements from the software requirements specifications (SRS) documents [MDS16]. RS and SRS requirements can be traced to even higher-level specifications such as the ones describing general AUTOSAR features and AUTOSAR's objectives. However, we advise AUTOSAR beginners to stick to the TPS and SWS specifications, at least at the beginning, as they are the ones that contain explanations and diagrams needed for understanding the AUTOSAR features in detail.

There are two additional general recommendations that we could give to readers who want to learn more about AUTOSAR. First, AUTOSAR specifications are not meant to be read from the beginning until the end. It is therefore recommended to switch between different specifications in a search for explanations related to a particular topic. Second, the readers should always read the latest AUTOSAR specifications as they contain up-to-date information about the current features of the AUTOSAR standard.

Apart from the specifications released by AUTOSAR, readers interested in knowing more about the AUTOSAR standard could find useful information in a few scientific papers. Related to the AUTOSAR methodology, Briciu et al. [BFH13] and Sung et al. [SH13] show an example of how AUTOSAR software components shall be designed according to AUTOSAR and Boss et al. [Bos12] explain in more detail the exchange of artifacts between different roles in the AUTOSAR development process, e.g., OEMs and Tier1s.

Related to the AUTOSAR meta-model, Durisic et al. [DSTH16] analyze the organization of the AUTOSAR meta-model and show possible ways in which it could be re worked in order to be compliant with the theoretical meta-modeling concept of strict meta-modeling. Additionally, Pagel et al. [PB06] provide more details about the generation of the AUTOSAR's XML schema from the AUTOSAR meta-model, and Brorkens et al. [BK07] show the benefits of using XML as an AUTOSAR exchange format.

Related to the configuration of AUTOSAR basic software, Lee et al. [LH09] explain further the use of the AUTOSAR meta-model for the configuration of AUTOSAR basic software modules. Finally, Mjeda et al. [MLW07] connect the phases of automotive architectural design based on AUTOSAR and functional implementation of the AUTOSAR software component in Simulink.

4.9 Summary

Since its beginning in 2003, AUTOSAR soon became a world-wide standard in the development of automotive software architectures, accepted by most major car manufacturers in the world. In this chapter, we explained the reference layered system architecture defined by AUTOSAR that is instantiated in dozens of car ECUs, and how different architectural components are usually developed according to the AUTOSAR methodology. We showed the role of the AUTOSAR meta-model in the design of the architectural components and the exchange of architectural models between different parties in the automotive development process. We also described major components of the AUTOSAR middleware layer (basic software) and how they could be configured.

Towards the end of the chapter, we visualized the evolution of the AUTOSAR standard by analyzing its meta-model and requirements changes between the latest AUTOSAR releases, and showed that the standard is still growing by standardizing new features. We also showed how AUTOSAR plans to support future cars functionalities, such as autonomous drive and car-to-car communication, by presenting ideas behind the AUTOSAR adaptive platform.

In our future work we plan to further analyze the differences between AUTOSAR classic and adaptive platforms on the meta-model and requirements levels, and the impact of using both platforms in the design of the automotive software systems.

References

AK03.   C. Atkinson and T. Kühne. Model-Driven Development: A Metamodeling Foundation. *Journal of IEEE Software*, 20(5):36–41, 2003.

AUT13.   AUTOSAR, http://www.nhtsa.gov. *National Highway Traffic Safety Administration*, 2013.

AUT16a.   AUTOSAR, www.autosar.org. *Automotive Open System Architecture*, 2016.

AUT16b.   AUTOSAR, www.autosar.org. *AUTOSAR Adaptive Platform for Connected and Autonomous Vehicles*, 2016.

AUT16c.   AUTOSAR, www.autosar.org. *Diagnostic Communication Manager v4.2.2*, 2016.

AUT16d.   AUTOSAR, www.autosar.org. *Diagnostic Event Manager v4.2.2*, 2016.

AUT16e.   AUTOSAR, www.autosar.org. *General Requirements on Basic Software Modules v4.2.2*, 2016.

AUT16f.   AUTOSAR, www.autosar.org. *Generic Structure Template v4.2.2*, 2016.

AUT16g.   AUTOSAR, www.autosar.org. *Layered Software Architecture v4.2.1*, 2016.

AUT16h.   AUTOSAR, www.autosar.org. *Methodology Template v4.2.2*, 2016.

AUT16i.   AUTOSAR, www.autosar.org. *Release Overview and Revision History v4.2.2*, 2016.

AUT16j.   AUTOSAR, www.autosar.org. *Software Component Template v4.2.2*, 2016.

BFH13.   C. Briciu, I. Filip, and F. Heininger. A New Trend in Automotive Software: AUTOSAR Concept. In *Proceedings of the International Symposium on Applied Computational Intelligence and Informatics*, pages 251–256, 2013.

BG01.   Jean Bézivin and Olivier Gerbé. Towards a Precise Definition of the OMG/MDA Framework. In *International Conference on Automated Software Engineering*, pages 273–280, 2001.

BK07.   M. Brörkens and M. Köster. Improving the Interoperability of Automotive Tools by
            Raising the Abstraction from Legacy XML Formats to Standardized Metamodels. In
            *Proceedings of the European Conference on Model Driven Architecture-Foundations
            and Applications*, pages 59–67, 2007.
BKPS07.   M. Broy, I. Kruger, A. Pretschner, and C. Salzmann. Engineering Automotive
            Software. In *Proceedings of the IEEE*, volume 95 of 2, 2007.
Bos12.   B. Boss. Architectural Aspects of Software Sharing and Standardization: AUTOSAR
            for Automotive Domain. In *Proceedings of the International Workshop on Software
            Engineering for Embedded Systems*, pages 9–15, 2012.
DST15.   D. Durisic, M. Staron, and M. Tichy. ARCA - Automated Analysis of AUTOSAR
            Meta-Model Changes. In *International Workshop on Modelling in Software Engineer-
            ing*, pages 30–35, 2015.
DSTH14.   D. Durisic, M. Staron, M. Tichy, and J. Hansson. Evolution of Long-Term Industrial
            Meta-Models - A Case Study of AUTOSAR. In *Euromicro Conference on Software
            Engineering and Advanced Applications*, pages 141–148, 2014.
DSTH16.   D. Durisic, M. Staron, M. Tichy, and J. Hansson. Addressing the Need for Strict Meta-
            Modeling in Practice - A Case Study of AUTOSAR. In *International Conference on
            Model-Driven Engineering and Software Development*, 2016.
ETS16.   ETSI, www.etsi.org. *Intelligent Transport Systems*, 2016.
Gou10.   P. Gouriet.   Involving AUTOSAR Rules for Mechatronic System Design.   In
            International Conference on Complex Systems Design & Management, pages 305–
            316, 2010.
Kru95.   P. Kruchten. Architectural Blueprints - The "4+1" View Model of Software Architec-
            ture. *IEEE Softwar*, 12(6):42–50, 1995.
Küh06.   T. Kühne. Matters of (Meta-) Modeling. *Journal of Software and Systems Modeling*,
            5(4):369–385, 2006.
LH09.   J. C. Lee and T. M. Han. ECU Configuration Framework Based on AUTOSAR ECU
            Configuration Metamodel. In *International Conference on Convergence and Hybrid
            Information Technology*, pages 260–263, 2009.
LLZ13.   Y. Liu, Y. Q. Li, and R. K. Zhuang. The Application of Automatic Code Generation
            Technology in the Development of the Automotive Electronics Software. In *Inter-
            national Conference on Mechatronics and Industrial Informatics Conference*, volume
            321–324, pages 1574–1577, 2013.
MDS16.   C. Motta, D. Durisic, and M. Staron. Should We Adopt a New Version of a Standard?
            - A Method and its Evaluation on AUTOSAR. In *International Conference on Product
            Software Development and Process Improvement*, 2016.
MLW07.   A. Mjeda, G. Leen, and E. Walsh. The AUTOSAR Standard - The Experience of
            Applying Simulink According to its Requirements. *SAE Technical Paper*, 2007.
NDWK99.   G. Nordstrom, B. Dawant, D. M. Wilkes, and G. Karsai. Metamodeling - Rapid Design
            and Evolution of Domain-Specific Modeling Environments. In *IEEE Conference on
            Engineering of Computer Based Systems*, pages 68–74, 1999.
Obj04.   Object Management Group, www.omg.org. *MOF 2.0 Core Specification*, 2004.
Obj14.   Object Management Group, http://www.omg.org/mda/. *MDA guide 2.0*, 2014.
PB06.   M. Pagel and M. Brörkens. Definition and Generation of Data Exchange Formats in
            AUTOSAR. In *European Conference on Model Driven Architecture-Foundations and
            Applications*, pages 52–65, 2006.
SH13.   K. Sung and T. Han. Development Process for AUTOSAR-based Embedded System.
            Journal of Control and Automation, 6(4):29–37, 2013.
Völ13.   L. Völker. SOME/IP - Die Middleware für Ethernet-basierte Kommunikation. *Hanser
            automotive networks*, 2013.

Chapter 5
Detailed Design of Automotive Software

Abstract Having discussed architectural styles and one of the major standards impacting architectural design of automotive software systems, we can now discuss the next abstraction level—detailed design. In this chapter we continue to dive into the technical aspects of automotive software architectures and we describe ways of working when designing software within particular software components. We present methods for modelling functions using Simulink modelling and we show how these methods are used in the automotive industry. We dive deeper into the need for modelling of software systems with Simulink by presenting an example of the braking algorithm and its implementation in Simulink (the example can be extended by the Simulink tutorials from Matlab.com). After presenting the most common design method—Simulink modelling—we discuss the principles of design of safety-critical systems in C/C++. We also introduce the MISRA standard, which is a standard for documenting and structuring C/C++ code in safety-critical systems.

5.1 Introduction

Architecting and high-level description of the automotive car software is usually the domain of OEMs. They decide what they want in their cars and what requirements they pose on their software system and electrical system. OEMs are responsible for breaking requirements on the system level to requirements on particular software components.

However, detailed design of software components and their subsequent implementation is the domain of suppliers (both Tier-1, Tier-2 and Tier-3) or in-house software development teams. It is these suppliers and in-house development teams that understand the requirements of the components, design the architecture of the components, implement the software, integrate it and then test it before delivering to the OEMs.

In this chapter we go through the principles of detailed design of automotive software. We start by describing the method used widely—Simulink modelling—then move to the principles of programming of safety-critical embedded systems and finally discuss principles of good programming according to the MISRA standard.

© Springer International Publishing AG 2017
M. Staron, *Automotive Software Architectures*,
DOI 10.1007/978-3-319-58610-6_5

5.2 Simulink Modelling

The models used in the design of automotive software often reflect the behavior of the function of a car and therefore, as such, are created in formalisms which reflect the physical world rather than the software world.

This kind of designing has implications on the design process and the competence of the designers. The process is shown in Fig. 5.1.

First of all, the process starts by describing the function of a car as a mathematical function in terms of its input, and outputs, with the focus on data flow. This means that the designers often operate with mathematical models to describe the function of a car. For instance, in order to describe the Anti-lock Breaking System (ABS, a well-known example from Matlab/Simulink), the designers need to describe the physical processes of wheel slippage, torque and velocity as a function (or functions) of time. When the mathematical descriptions are ready, each of the equations is translated to a set of Simulink blocks.

Fig. 5.1 Designing using Simulink models—a conceptual overview

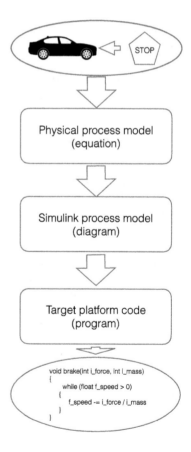

When translating the mathematical equations into Simulink blocks, transitions and functions, the designers focus on the flow of the data and the feedback loops present there. For example, in the ABS example the slippage of the wheel depends on the speed and the speed depends on the slippage. These feedback loops are present in the model. In more advanced cases, the designers need to write pieces of code in Matlab to describe some of the functions which are not available in the standard Simulink libraries.

Once the model is completed and tested, it is used to generate the code in the target programming language—usually C or C++, depending on the system.

In this section we go into more depth about this process.

5.2.1 Basics of Simulink

Simulink has a rich library of functions and blocks which help the designers to model their systems. We present the main blocks and describe their usage.

The basic principle of each Simulink model is that it starts with the source and ends with a sink, which means that there is data flowing through a number of steps in the process, starting from the source and ending in the sink.

The usual sources of the data are the function blocks or the step blocks. The usual sinks in the model are either scope blocks (for observing the outcome) or the output ports of the models.

5.2.1.1 Function and Step Blocks

The model usually "starts" with the step block, which provides the basic input to the entire model and allows for its simulation. The standard source blocks are shown in Fig. 5.2. The figure shows only a subset of the most commonly used source blocks in automotive software design.

The meaning of these blocks is:

- Constant—generates the signal of a constant value.
- Clock—generates the signal which is the current time of the simulation.
- Digital clock—generates the simulation signal at specific periods of time.
- Pulse generator—generates a pulse, where all parameters can be specified.
- Ramp—generates a signal which is constantly increasing or decreasing at a specified rate.
- Random number—generates a random number for the simulation.
- Signal generator—generates some of the most commonly used signals such as a Sine wave or a specific function.
- Step—generates a discrete step signal of which the value and frequency can be specified.

Fig. 5.2 Simulink basic
blocks—sources of signals in
the Simulink model

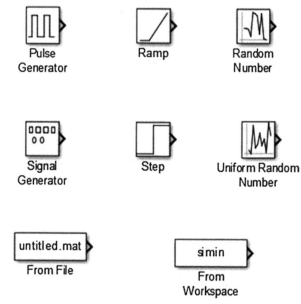

- Uniform random number—generates a random number which is evenly distributed over a specified interval.
- From file—generates a set of signals which are stored in a file (which can be the result of simulations from other models).
- From workspace—similar to from a file, but with signals that do not store the time.

The source blocks can provide the input signals in terms of continuous signals (e.g. a Sine wave), discrete signals (e.g. Step blocks), random signals (e.g. random number) or a predefined sequence (e.g. from File).

5.2.1.2 Commonly Used Blocks

The blocks which are collected under the category of the most commonly used blocks are:

- Gain—gives the output as a multiplication of the input (the multiplier is specified by the designer).
- Product—gives the output as the product of two inputs (e.g. signals).
- Sum—similar to the Product block, but shows the output as the sum of two signals.
- Saturation—imposes upper and lower limits on the input signal.

- Subsystem—a block representing a subsystem (e.g. an embedded model). This type of block is used very often to structure models into hierarchies and to use one model as part of another one.
- Out1—models a signal that goes outside of the current model (e.g. to another model).
- In1—the opposite to Out1—used to take the signal from outside of the current model into the simulation.
- Integrator—where the output is the integral of the input.
- Switch—a block which chooses between the first and the third input based on the value of the second input.
- Terminator—a block used to capture signals which do not connect to other blocks.

The graphical symbols for these blocks are shown in Fig. 5.3.

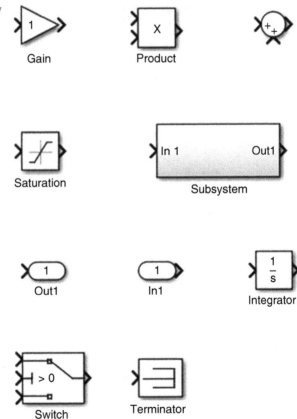

Fig. 5.3 Simulink commonly used blocks

5.2.1.3 Sinks

The standard blocks that are used as sinks of the models are:

- Display—the current value of the step of the simulation at specific location of the simulation.
- Scope—diagram showing the display as a function of time of the simulation.
- Stop—stopping the simulation when the signal is other than zero.
- To file—sending the signal to the specified file.
- To workspace—storing the signal without the time variable.
- XY graph—diagram used to plot two signals against each other (instead of against time).

The graphical representation of these blocks is presented in Fig. 5.4.

In the design of physical processes it is often the case that we need to describe a process as a mathematical function. The Matlab environment is well suited for that purpose and the Simulink environment can take advantage of all built-in and user-defined functions. The basic block used for that is presented in Fig. 5.5.

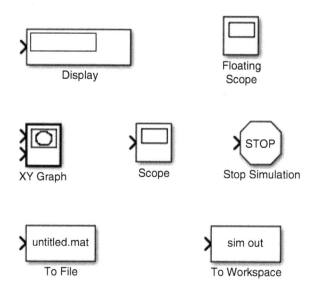

Fig. 5.4 Simulink model sink blocks

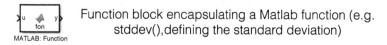

Function block encapsulating a Matlab function (e.g. stddev(),defining the standard deviation)

Fig. 5.5 Simulink basic blocks—Matlab function encapsulation in the Simulink model

5.2.2 Sample Model of Digitalization of a Signal

Let us now focus on designing a simple Simulink model which converts an analog signal to a digital one. This process can be described in Formula (5.1).

$$f(x) = \begin{cases} 1 & if\ x > 0 \\ 0 & if\ x \le 0 \end{cases} \tag{5.1}$$

This equation corresponds to the Simulink model presented in Fig. 5.6.

The equation is specified in the middle block—the "Compare to constant" block named "Digitalize signal", as shown in Fig. 5.7.

The main part of the figure is the two options—Operator and Constant. They are also shown in the icon in Fig. 5.6.

Now that we have the digitalization function, we need to package that into a block with two ports—input and output. We can also add an example function that will generate a signal used to test the block—as presented in Fig. 5.8.

Figure 5.8 shows three blocks: The sine wave function (left-hand side) generates the signal to digitalize; the scope block (right-hand side) is used to visualize the results of the simulation. The scope block has two inputs—one from the sine

Fig. 5.6 Digitalization of a signal value as designed in Simulink

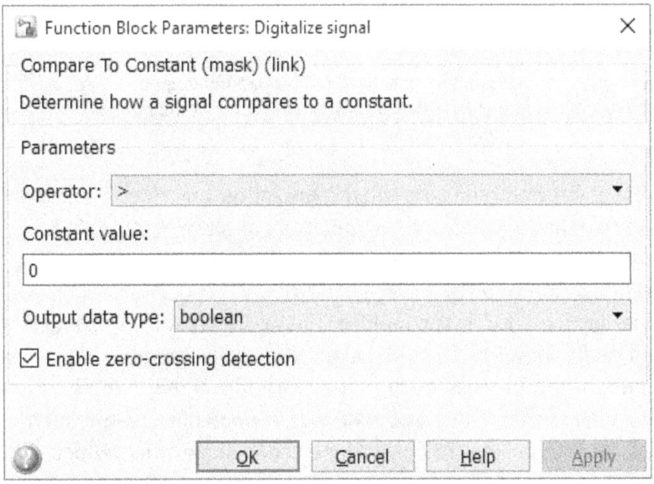

Function Block Parameters: Digitalize signal	×
Compare To Constant (mask) (link)	
Determine how a signal compares to a constant.	
Parameters	
Operator: >	▼
Constant value:	
0	
Output data type: boolean	▼
☑ Enable zero-crossing detection	
	OK Cancel Help Apply

Fig. 5.7 Specification of the digitalization function in the Digitalize signal block

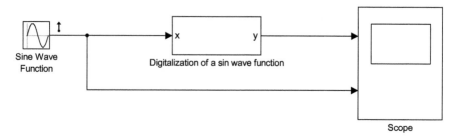

Fig. 5.8 Making the digitalization into a Simulink block

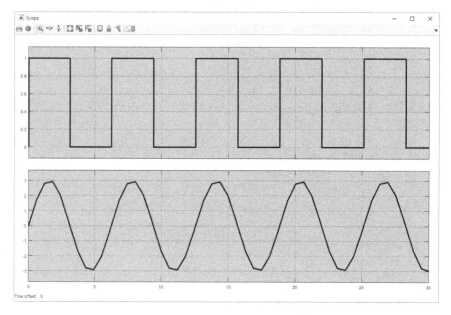

Fig. 5.9 The result of the simulation visualized as two parallel diagrams—the digitalized result at the *top* and the original input as provided by sine wave source at the *bottom*

wave function itself and one from the digitalization function. These two inputs are visualized in two diagrams after the simulation, as shown in Fig. 5.9.

The newly designed block contains the diagram presented in Fig. 5.6 and is named "Digitalization of a sin wave function".

The model presented in this example is naturally very simple and illustrates the simplicity of using Simulink to model a mathematical equation. Now, this particular equation is about the process of digitalization of a signal, which is not based on physical processes in real life. The model also does not contain such elements as the feedback loop important in designing of control systems (which we expand on in the upcoming sections).

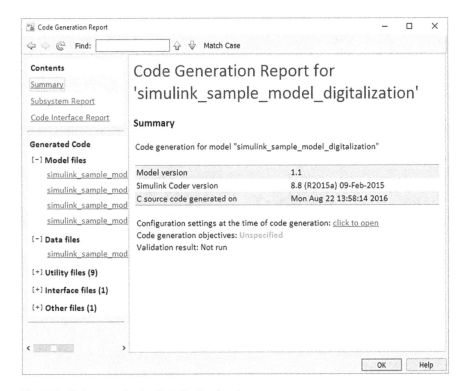

Fig. 5.10 Code report for the digitalization function

The next step in the design of a system based on the digitalization block is to generate the C/C++ code from the model. The code generated from this Simulink model has the property that it is hard to read for a human being and therefore the Simulink environment provides a report about what has been generated. The report for this model is presented in Fig. 5.10.

The report guides us to all the files that were generated ("Model files" in the left-hand side of the figure) and provides the summary in the main window.

The actual piece of code can look like the code presented in Fig. 5.11. The code in the figure presents a C structure with the initialization of the blocks (e.g. the sine wave parameters and the digitalization threshold "0").

5.2.2.1   Comments on the Sample Model

In this simplistic example we managed to see the power of Simulink and at the same time we managed to follow the process of designing automotive software as shown in Fig. 5.1. In the design of the automotive software we have libraries which take care of this kind of process. These libraries, however, are part of the lowest layers of automotive software and can be seen in the architecture diagram of a communication layer in the CAN bus communication.

```
21  #include "simulink_sample_model_digitalization.h"
22  #include "simulink_sample_model_digitalization_private.h"
23
24  /* Block parameters (auto storage) */
25  P_simulink_sample_model_digit_T simulink_sample_model_digital_P = {
26    0.0,                         /* Mask Parameter: Digitalizesignal_const
27                                  * Referenced by: '<S2>/Constant'
28                                  */
29    3.0,                         /* Expression: 3
30                                  * Referenced by: '<Root>/Sine Wave Function'
31                                  */
32    0.0,                         /* Expression: 0
33                                  * Referenced by: '<Root>/Sine Wave Function'
34                                  */
35    1.0,                         /* Expression: 1
36                                  * Referenced by: '<Root>/Sine Wave Function'
37                                  */
38    0.0                          /* Expression: 0
39                                  * Referenced by: '<Root>/Sine Wave Function'
40                                  */
41  };
42
```

Fig. 5.11 Generated source code for the initialization of the block

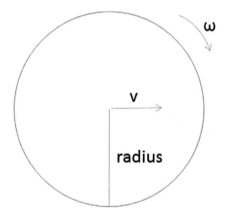

Fig. 5.12 Relation between linear and wheel velocity

5.2.3 *Translating Physical Processes to Simulink*

The example with the digitalization of the signal is rather trivial and has no physical process that is modelled. However, in most cases of Simulink modelling in automotive software, we have such models.

To illustrate that such processes are modelled both as mathematical equations and as Simulink blocks, let us consider an example of calculating the linear velocity of a wheel based on its wheel velocity and vice versa. Figure 5.12 shows the relation between these two kinds of velocity for a wheel of radius "radius".

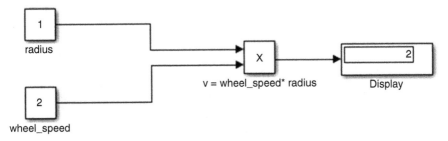

Fig. 5.13 Simulink model calculating the linear velocity

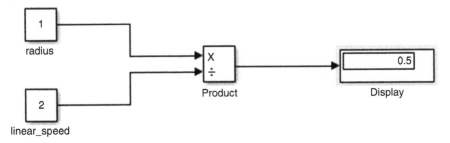

Fig. 5.14 Simulink model calculating the wheel velocity

The equations describing the relation between the two velocities are:

$$v = \omega * radius \tag{5.2}$$

and

$$\omega = \frac{v}{radius} \tag{5.3}$$

Both of the equations are rather simple and let us now build the model which will take two scalar values and calculate the linear velocity. The model is presented in Fig. 5.13.

The model consists of two scalar values (wheel speed and radius), their product and the display sink. Executing the model displays the result in the display sink.

The model which calculates wheel velocity based on linear velocity requires changing the product to be a fraction instead. The resulting model is presented in Fig. 5.14.

The properties of the product block are changed as presented in Fig. 5.15.

In the "Number of inputs" field we make a change, which denotes division instead of multiplication.

Before we move to another example, let us illustrate another important concept in the design of control systems using Simulink—feedback loops. The concept of a

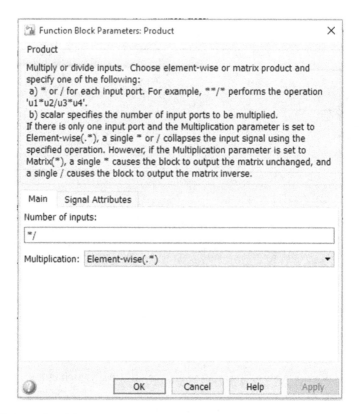

Fig. 5.15 Properties of the product block

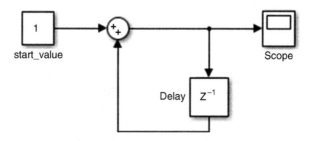

Fig. 5.16 Example of a simple feedback loop

feedback loop is often used in control systems to design self-regulating systems. In Fig. 5.16 we can see an example of a simple feedback loop.

In the figure we can see that the loop takes a signal directly from the output of the summation and puts it back with a delay. The delay is needed in order to make sure that the first iteration of the simulation has the initial value in the summation. The properties of the delay block are shown in Fig. 5.17.

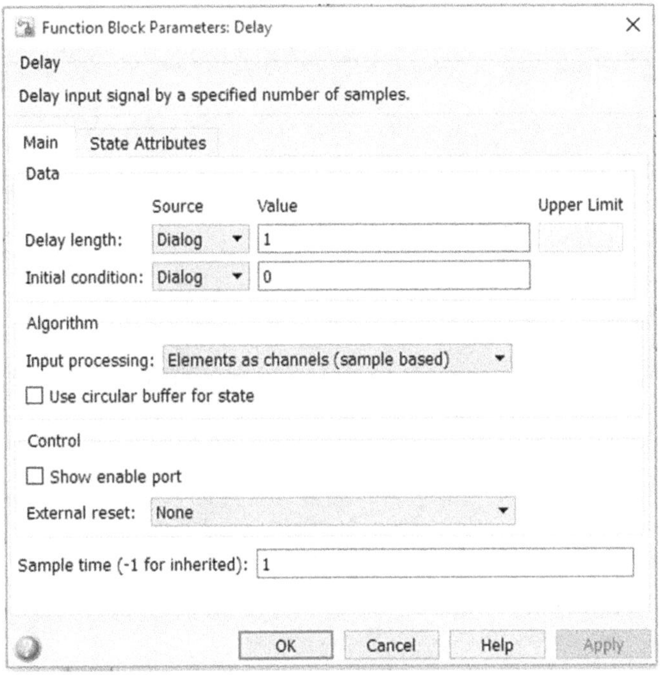

Fig. 5.17 Properties of the delay block

The important part is the "Delay length" property, which denotes how many simulation cycles the input signal is postponed. Once we execute the simulation, we can see that the summation results in the gradual increase of the signal as shown in Fig. 5.18.

5.2.4 Sample Model of Car's Interior Heater

Now let us look into a bit more complex model—the heater of a car. The model introduces the feedback loop and has been inspired by the house heating model from the Matlab Simulink standard model library, but has been simplified to illustrate only the most important aspects of modelling systems with the control loop.

In general the model of a heater contains three components, which we will turn into blocks:

- Car interior—describing the temperature of the car's interior, including heat loss
- Heater—describing the heater, its on/off status and the heating temperature
- Thermostat—describing the switch for the heater

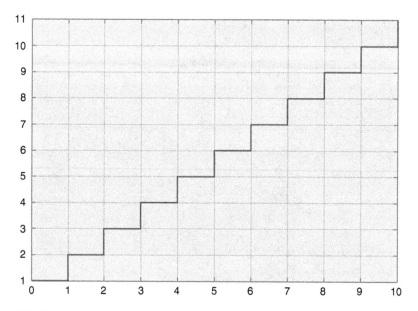

Fig. 5.18 Results of the simulated feedback result

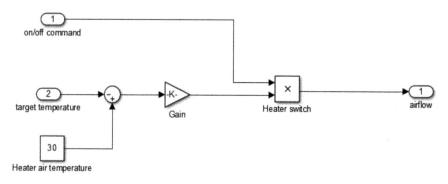

Fig. 5.19 Heater model

There are two inputs to this simulation model—the outdoor temperature and the desired temperature of the interior.

Let us start with modelling the heater itself. The heater has an on/off switch for the flow of the air as well as the heater element. This means that when it is switched on, it blows hot air at a given temperature into the interior compartment of the modelled car. A simple model can look as in Fig. 5.19.

In this model the heater blows hot air of temperature 30 °C at a given rate (modelled as Gain K). The gain block is configured as shown in Fig. 5.20.

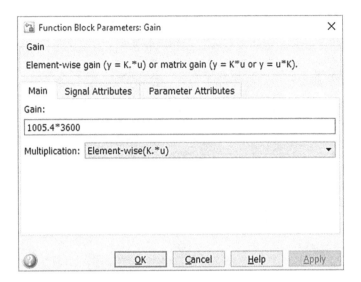

Fig. 5.20 Heater model—gain block properties

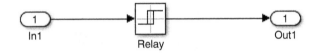

Fig. 5.21 Switcher model

The two constants that are multiplied by each other are (1) the air flow per hour, which we assume is a constant rate of 1 kg/s, which gives 3600 ks/h, and (2) the heat capacity of the air, which is 1005.4 J/kg-K at the room temperature (in our model).

We need to make a small observation here—the values which we use in the model are mostly constant, as we want to illustrate how to design an algorithm in Simulink. However, in real life the challenge is to model these constants as functions. For example, we assume the heat capacity of air to be constant, which is not accurate as it changes with the temperature of the air. The flow rate of the heater is also not constant as when the heater starts the fan needs some time to start spinning and therefore the flow rate changes. In reality we could have two equations modelling these two processes and use them as input instead of providing constants.

Now, let us move over to the model of the switch of the heater, which needs to switch on and off the heater based on the difference in the temperature outside of the car. Let us configure the on/off deviation to be 3 °C compared to the desired temperature. We can use the relay block to model that, as shown in Fig. 5.21.

The properties of the relay are the on/off criteria $(+/-3°)$ and the output signal for on (1) and off (0), as shown in Fig. 5.22.

The next step is to link both blocks together as shown in Fig. 5.23. The link has to connect the input on/off port of the heater to the output on/off port of the switcher.

Function Block Parameters: Relay   ✕

Relay

Output the specified 'on' or 'off' value by comparing the input to the
specified thresholds. The on/off state of the relay is not affected by input
between the upper and lower limits.

Main   Signal Attributes

Switch on point:

| ⊟ |

Switch off point:

| -3 |

Output when on:

| 1 |

Output when off:

| 0 |

Input processing: [Elements as channels (sample based)   ▼]

☑ Enable zero-crossing detection

| | OK | | Cancel | | Help | | Apply |

Fig. 5.22 Switcher model—relay properties

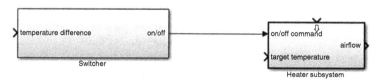

Fig. 5.23 Linking the heater to the switcher

Now, we need to model the environment and the feedback loops before we go
into modelling the car's interior. In particular we need to model the calculation of
the temperature difference between the interior and the desired temperature. We do
it by adding a proxy for the car (an empty subsystem), which we will design in the
next steps by adding the summation component to calculate the difference between
the desired and the current temperature. We also need to add a constant which
configures the model with the desired temperature. We do it by adding a constant
block and setting the temperature to 21 °C. The resulting model is presented in
Fig. 5.24.

The model has one port which is not connected—it is the current temperature
port of the heater; we need to connect this to a signal from the interior of the car.

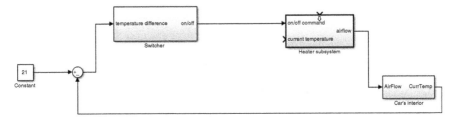

Fig. 5.24 First version of the air heater model with the feedback loop

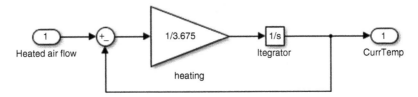

Fig. 5.25 Model of the interior of the car

Now, we need to model the actual temperature of the car's interior. The temperature of the car's interior is the same as the temperature outside (which we need to add to our model) and increases as the heater blows in the hot air. The increase of the temperature of the interior can be described by the following equation:

$$\frac{dTemp_{car}}{dt} = \frac{1}{M_{air} * 1005.4 J/kg - K} * \left(\frac{dQ_{heater}}{dt}\right) \tag{5.4}$$

Now, for a normal car, the mass of the air (M_{air}) is a product of the volume of the car's interior and the density of the air (a constant of $1.2250 \, kg/m^3$). In order to simplify things, let's say that the volume of a personal vehicle's interior is $3 \, m^3$, which, multiplied by the density of the air gives $3.675 \, kg$ as the mass of the air. Now we have a model which looks like the one in Fig. 5.25.

In the model we use the gain block to increase the temperature and the integrator to set the initial temperature. We also add the feedback loop to make the increase in the temperature similar to a loop in a programming language. Inside the gain block we put the calculated increase in temperature as shown in Eq. (5.4), resulting in the configuration shown in Fig. 5.26.

When we connect all the elements, we get the following model—Fig. 5.27.

When we look at the plot of the temperature over time we can see the result as shown in Fig. 5.28.

Now we can see that the model is too simplistic. The temperature of the car's interior goes up from the initial value of 1 and then stays at the constant level. It is because our model of the car's interior takes into consideration only the heating process of the interior, at the same time ignoring the process of chilling the interior

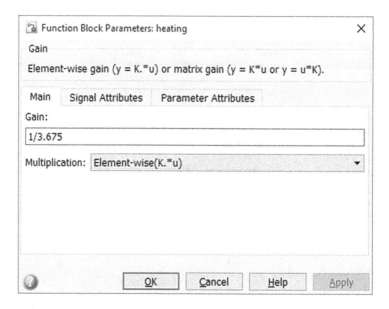

Fig. 5.26 Properties of the gain block in the interior

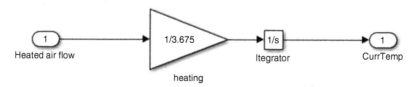

Fig. 5.27 Heating-only model of the car's heater

when the heater is not working. In order to fix that without complicating the model too much, let us add a feedback loop after the gain block, in the way shown in Fig. 5.29.

Once we make this addition, we can see that the temperature of the car's interior drops when the heater is not powered on, as shown in Fig. 5.30.

5.2.4.1 Summary of the Heater Model

The heater model presented in this section is a simplistic model with a feedback loop and illustrates a few important principles which make Simulink modelling so popular in software development.

Once the model is somewhat complete, the designers can execute the model and observe the results. As the "Scope" sink can be placed virtually at any signal, it is easy to debug the models and to understand where it does not work (if needed).

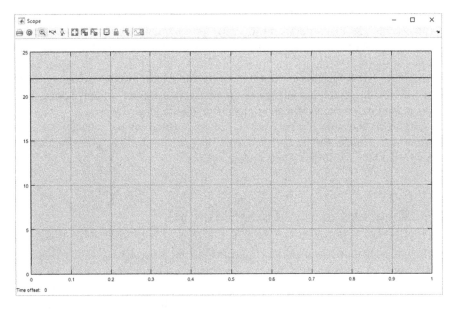

Fig. 5.28 Result of a simulation of the heater model

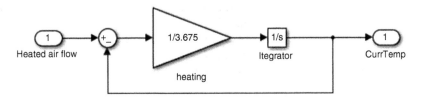

Fig. 5.29 Model of the interior with cooling effect

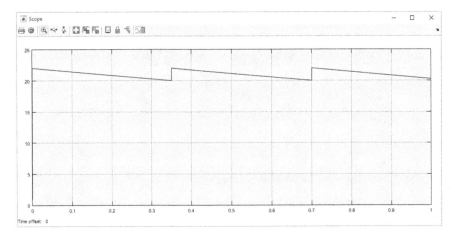

Fig. 5.30 Result of the simulation with the cooling effect

Another principle is the ability to make the model modular. The designers can use constants and assumptions during early prototyping phases of their software development. As the development progresses and the designers know more about the physical processes, they can replace constants with calculations of values using blocks and Matlab functions. These functions can be developed either analytically based on the designer's knowledge of the physical processes or they can be done using mathematical regression and statistical modelling techniques.

And finally the ability to generate source code which can be executed on target platforms. If a model can be executed, then the code for it can be generated, which is a very big help for automotive software engineers.

5.3 Simulink Compared to SySML/UML

SySML is a notation based on the Unified Modelling Language (UML). Compared to the Simulink notation, it is different and neither of them has a specific software development process which the notation supports. However, in practice these two notations support different development processes. In Fig. 5.31 we outline these differences per phase of software development.

In the **analysis** phase these two notations support different types of analysis and modelling. Simulink is based on describing the system using mathematical equations (as we saw in the examples in this chapter) whereas SySML/UML use conceptual models and class diagrams (with low level of detail). The models created in SySML/UML are intended to be high level and non-executable whereas the mathematical models need to be rather complete as they will be used in modelling in the **design** phase.

In the design phase the main goal is to develop a detailed model of the software and there these two notations differ significantly. In SySML/UML the main entities

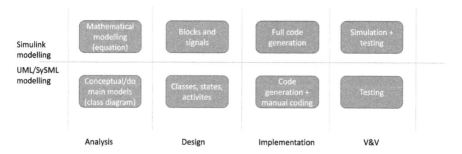

Fig. 5.31 Result of the simulation with the cooling effect

are classes (corresponding to programming language classes/modules), statecharts and sequence diagrams. Although the SySML/UML notations provide more types of diagrams than these three, these three are by far the most popular ones. In Simulink the primary entities are blocks and signals, as we saw in the examples in this chapter.

The **implementations** of the two designs differ significantly—Simulink usually results in 100% code generation. The generated code can be compiled and executed. The SySML/UML notations usually do not result in full code generation, but in so-called skeleton code. The skeleton code needs to be complemented by the designers with manually written code in the target programming language.

Once the designs are implemented, they are **tested**, which in Simulink happens through simulations (sometimes using test environments to execute the simulations), whereas for the SySML/UML generated code, the code is tested in a traditional manner, e.g. using unit tests.

The SySML/UML languages are often called architectural languages because they come from the field of object-oriented analysis and design and focus on the conceptual modelling of objects in the real world. This means that the main part of the effort is on the development of the design models, because all details of the target programming language have to be taken into consideration—otherwise we cannot generate the code. Therefore we can see that in the automotive domain these languages are often used to specify logical component architectures, whereas the detailed design of automotive systems is done using Simulink.

5.4 Principles of Programming of Embedded Safety-Critical Systems

Safety-critical systems have entered the automotive industry quite recently compared to the aviation and space industry [Sto96, Kni02]. Historically, the aviation industry and the space industry relied on the Ada programming language due to its well-defined semantics and mechanisms for parallel programming.

In the telecommunication industry engineers use functional programming languages such as Haskell or Erlang, even if the safety criticality is not that crucial there.

In the automotive industry, however, it is the generated C/C++ code which is the most common. C/C++ have the advantage of being relatively well known by the software engineering community, relatively simple if needed and with good compiler support. In practice this means that the code can be ported easily between different operating systems, as the majority of the safety critical OSs have the Unix kernel at their core.

The operating systems often used in automotive software are VxWorks and QNX, which are relatively simple, with great schedulers and task handlers. It is their simplicity that allows the designers to retain a large degree of control over the programs and therefore makes them so popular. The AUTOSAR standard standardizes a number of elements of the underlying operating systems, as discussed in Chap. 4.

As the software system of the car is distributed over multiple ECUs, it is the communication between the ECUs which is important. From the designer's perspective this communication means that there are signals exchanged between different software components and state machines need to be synchronized. Often, in the programming language this means that the messages are packaged as packages or sent using sockets.

From the physical perspective the designers have a number of different communication protocols available, such as:

- CAN bus—Specified in the ISO standard [Sta93], it is currently the most frequently used bus in the automotive industry. It allows us to send messages with a speed of up to 1 MBps, which allows to send even video streams in the car's bus (e.g. from the parking camera). The standard is popular because of its relatively simple architecture and specifications of the MAU (Medium Access Unit) and the DLL (Data Link Layer) parts.
- Flexray bus—Specified in the ISO 17458 standard, is one of the possible future directions of development in the automotive industry. It allows communications with a speed of up to 10 Mbps over a similar type of wiring and has two independent data channels (one for fault tolerance).
- Ethernet bus—Used throughout the internet for communications, it is now being considered for speeds of up to 1 Gbps. At the time of writing of this book the protocol is used for downloading new software to ECUs for many car manufacturers and for communications during driving some cars. As the protocol is prone to electrostatic distortions, the majority of the manufacturers are waiting for more mature specifications before they start using this protocol more widely in their electrical systems.
- MOST bus—Used in the automotive industry for sending/receiving multimedia-related content (e.g. video and audio signals). The communication speeds are up to 25–150 Mbps depending on the version of the standard.
- LIN bus—used for low cost communications with speeds of up to 20 Kbps between mechatronic nodes in the car.

In the design of the automotive systems, the architects usually decide upon the topology of the network and its communication buses rather early. As we can see from the description of each of these protocols, they are aimed at different purposes and therefore their choice is rather straightforward.

5.5   MISRA

When designing the software for automotive applications, we need to follow certain design guidelines. The automotive industry has adopted the MISRA-C standard where the details of the design of computer programs are in the C programming language [A+08]. The standard contains the principle of how to document embedded C code—in terms of naming conventions, documentation and the use of certain programming constructs. The rules are grouped into such categories as:

1. Environment—rules related to the programming environment used in the development (e.g. mixing of different compilers).
2. Language extension—rules specifying which types of comments are to be used, enclosing assembly code or removing commented code.
3. Documentation—rules defining which code constructs should be documented and how.
4. Character sets—usage of ISO C character sets and trigraphs.
5. Identifiers—defining the length and naming convention of identifiers as well as the usage of typedef.
6. Types—the usage of the "char" type, the naming convention of new types and the usage of bit fields.
7. Constants—preventing the usage of octal constants.
8. Declarations and definitions—rules about the explicit visibility of types of functions and their declarations.
9. Initialisation—rules about default values of variables at their declaration.
10. Arithmetic type conversions—describing implicit and explicit rules for type conversions as well as the dangerous conversions.
11. Pointer type conversions—rules regarding the interchangeability of different types of pointers.
12. Expressions—rules about the evaluation of arithmetical expressions in programs.
13. Control statement expressions—rules about the expressions used in for loops, explicit evaluations of values to Boolean (instead of 0).
14. Control flow—rules about the dead code, null statements and their location and prohibited goto statements.
15. Switch statements—rules about the structure of the switch statements (a subset of possible structures from the C language).
16. Functions—rules prohibiting such unsafe constructs as variable argument lists or recursion.
17. Pointers and arrays—rules about the usage of pointers and arrays.
18. Structures and unions—rules about the completeness of union declarations and their location in memory; prohibiting the usage of unions.
19. Preprocessing directives—rules about the usage of #include directives and C macros.
20. Standard libraries—rules about the allocation of heap variables, checking the parameters of library functions and prohibiting certain standard library functions/variables (e.g. errno).
21. Run-time failures—rules prescribing of usage of static analysis, dynamic analysis and explicit coding for avoiding runtime failures.

The MISRA rules are often encoded in the C/C++ compilers used in safety-critical systems. This inclusion in compilers makes it rather simple and straightforward and therefore widely used.

The MISRA standard was revised in 2008 and later in 2012, leading to the addition of more rules. Today, we have over 200 rules, with the majority of them classified as "required".

Let us now analyze one of the rules and its implications—we take rule #20.4: "Dynamic heap memory allocation shall not be used." This rule in practice prohibits dynamic memory allocations for the variables. The rationale behind this rule is the fact that dynamic memory allocations can lead to memory leaks, overflow errors and failures which occur randomly. Taking just the defects related to the memory leaks can be very difficult to trace and thus very costly. If left in the code, the memory leaks can cause undeterministic behavior and crashes of the software. These crashes might require restart of the node, which is impossible during the runtime of a safety-critical system. Following this rule, however, also means that there is a limit on the size of the data structures that can be used, and that the need for memory of the system is predetermined at design time, thus making the use of this software "safer".

5.6   NASA's Ten Principles of Safety-Critical Code

The United States-based NASA has a long tradition of developing and using safety-critical software. In fact, much of the initial reliability research has been done in the vicinity of NASA's Jet Propulsion Laboratory. The reason for that is that NASA's missions often require safety-critical software to steer their devices such as space shuttles or satellites.

In 2006 Holtzman presented ten rules of safety-critical programming, which come from NASA, but apply to all safety-critical software [Hol06]. These rules are (the original wording of the rules is kept):

1. Restrict all code to very simple control flow constructs, do not use goto statements, setjmp or longjmp constructs, direct or indirect recursion.
2. Give all loops a fixed upper bound. It must be trivially possible for a checking tool to prove statically that the loop cannot exceed a preset upper bound on the number of iterations. If a tool cannot prove the loop bound statically, the rule is considered violated.
3. Do not use dynamic memory allocation after initialization.
4. No function should be longer than what can be printed on a single sheet of paper in a standard format with one line per statement and one line per declaration. Typically, this means no more than about 60 lines of code per function.
5. The code's assertion density should average to minimally two assertions per function. Assertions must be used to check for anomalous conditions that should never happen in real-life executions. Assertions must be side effect-free and should be defined as Boolean tests. When an assertion fails, an explicit

recovery action must be taken, such as returning an error condition to the caller of the function that executes the failing assertion. Any assertion for which a static checking tool can prove that it can never fail or never hold violates this rule.

6. Declare all data objects at the smallest possible level of scope.
7. Each calling function must check the return value of non-void functions, and each called function must check the validity of all parameters provided by the caller.
8. The use of the preprocessor must be limited to the inclusion of header files and simple macro definitions. Token pasting, variable argument lists (ellipses), and recursive macro calls are not allowed. All macros must expand into complete syntactic units. The use of conditional compilation directives must be kept to a minimum.
9. The use of pointers must be restricted. Specifically, no more than one level of dereferencing should be used. Pointer dereference operations may not be hidden in macro definitions or inside typedef declarations. Function pointers are not permitted.
10. All code must be compiled, from the first day of development, with all compiler warnings enabled at the most pedantic setting available. All code must compile without warnings. All code must also be checked daily with at least one, but preferably more than one, strong static source code analyzer and should pass all analyses with zero warnings.

These rules are naturally captured by the MISRA rules and show the similarity of safety-critical systems regardless of the application domain. The "heart" of these rules is that the safety-critical should be simple and modularized. For example, the length of a typical function should be less than 60 lines of code (principle #4), which is supported by the limits of the maintainability of large and complex code.

What these principles also show is the difficulty of automatically checking for their violation. For example, the principles #6 ("Declare all data objects at the smallest possible level of scope") requires parsing of the code in order to establish the boundary of the "smallest possible level of scope").

5.7 Detailed Design of Non-safety-Critical Functionality

In the previous sections we focused on designing software which is often considered safety-critical to various extents. However, there is a significant amount of software in modern cars which is not safety-critical. One of such non-safety-critical domains is the infotainment domain, where the main focus is on connectivity and user experience of the interface. Let us look into one of the standards in this domain—GENIVI [All09, All14].

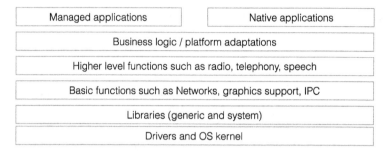

Fig. 5.32 GENIVI layered architecture overview

5.7.1 Infotainment Applications

The GENIVI standard is built upon a layered architecture with five basic layers, as shown in Fig. 5.32.

In the GENIVI architecture the top layers are designated to the user applications, which in turn can expose their services to one another. The standard itself, however, focuses on the basic and high-level functions [All15]. The following areas are included in the reference architecture:

- Persistence—providing persistent data storage
- Software management—supporting such functionality as SOTA (Software-Over-The-Air) updates
- Lifecycle—supporting the start-up and shutdown of the system
- User management—supporting multiple users and their profiles
- Housekeeping—supporting error management
- Security infrastructure—supporting cryptography and interactions with hardware security modules
- Diagnostics—supporting the diagnostics as specified in ISO 14229-1:2013
- Inter-Process Communications (IPC)—supporting communication between processes (e.g. message brokers)
- Networks—supports the implementation of different vehicle network technologies (e.g. CAN)
- Network management—supports the management of network connections
- Graphics support—providing graphics libraries
- Audio/Video processing—providing codecs for audio and video playback
- Audio management—supporting the streaming and prioritizing streams of audio
- Device management—providing support for devices via (for example) USB
- Bluetooth—providing the bluetooth communication stack
- Camera—providing the functionalities needed for vehicle cameras (e.g. rearview camera)
- Speech—supporting voice commands
- HMI support—provides the functionality to handle user interactions

- CE Device integration—supports such protocols as CarPlay
- Personal Information management—supporting the basic functionality of address book and passwords
- Vehicle interface—provides the possibility to communicate with other vehicle systems
- Internet functions—provides the support for internet, e.g. web browsing
- Media sources—provides support for media sharing such as DLNA
- Media framework—provides the generic logic of media players
- Navigation and Location Based Services—supporting the navigation systems
- Telephony—provides the support for telephony stack
- Radio and tuners—provides the support for radio

The above list shows that the GENIVI reference architecture is a large step towards standardization of the internals of infotainment systems, which will allow users to use common software ecosystems rather than OEM-specific solutions.

Today we can see the GENIVI implementation in many car platforms, such as BMW with the system from Magneti Marelli (according to the GENIVI website). The standard ADL for the GENIVI applications is the Franca IDL which is used for defining interfaces in GENIVI software components.

5.8 Quality Assurance of Safety-Critical Software

Quality assurance of automotive software follows a number of standards, one of them being the ISO/IEC 25000 series of standards [ISO16]. The usual way that the standards describe the quality is that they divide the quality into a set of characteristics and a set of perspectives. The three perspectives on software quality are:

1. External software quality—describing the quality of the software product in relation to its requirements (hence the classification as "external").
2. Internal software quality—describing the quality of the software in relation to the construction of the software (hence the classification as "internal").
3. Quality in use—describing the quality of the software from the perspective of its users (hence the classification as "in use").

In this chapter we focus on the internal quality of the software and the methods to monitor and control the internal quality—formal methods for verifying the correctness of the software and static analysis for verifying properties of software such as complexity. Testing as a technique for finding defects has been discussed in Chap. 3.

5.8.1 Formal Methods

Formal methods is a term used to collectively denote a set of techniques for specification, development and verification of software using formalisms related to mathematical logic, type theories, and symbolic type executions.

In the automotive domain formal methods are required during the verification of ASIL D components (classified according to the ISO/IEC 26262 standard; see Chap. 8).

The verification in the formal way often follows a strict process where the software is specified using a formal notation (e.g. a VDM) and then gradually refined into source code of the program. Each step is shown to be correct and therefore the software is formally proven to be correct.

5.8.2 Static Analysis

Another method for assuring the internal quality of automotive software is the static analysis [BV01, EM04]. Static analysis refers collectively to a set of techniques for analyzing the source code (or the model code) of a software system. The analysis aims at discovering vulnerabilities in the software code and violations of programming good practices. Static analysis in the automotive systems usually looks for violations of MISRA rules and good coding rules.

In addition to the MISRA rules, the static analysis often checks for the following (examples):

- API usage errors, for example, using of private APIs
- Integer handling issues, for example, potentially dangerous type casts
- Integer overflows during calculations
- Illegal memory accesses, for example, using of pointer operations
- Null pointer dereferences
- Concurrent data access violations
- Race conditions
- Security best practices violations
- Uninitialized members

In order to analyze a program statically no execution is needed and therefore this technique is very popular. The majority of static analysis tools do not need the code to actually execute and therefore there is no requirement for the code to be complete and runnable, which is the case for formal analysis (e.g. symbolic execution) or dynamic analysis.

An example screenshot from one of the tools for static analysis (SonarQube) is presented in Fig. 5.33.

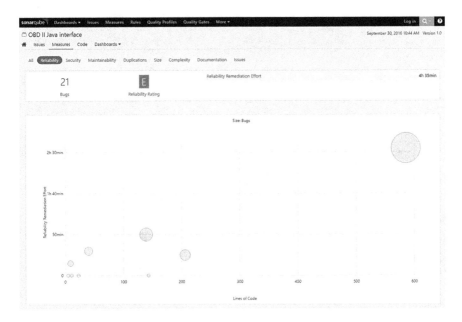

Fig. 5.33 Screenshot from SonarQube static analysis software

Fig. 5.34 Screenshot from SonarQube static analysis software, customized dashboard

In the figure we can see the development of complexity per module. The complexity has direct impact on testability (higher complexity, lower testability), and therefore it is an important parameter of the internal quality of the software.

Another view is presented in Fig. 5.34. The figure presents a custom view on the quality—complexity per class and percentage of duplicated (cloned) code.

SonarQube can be expanded with the help of plug-ins to include multiple programming languages and analyses; it can also be extended by custom plug-ins.

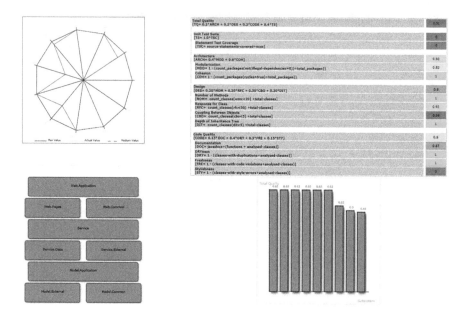

Fig. 5.35 Screenshot from XRadar static analysis software

However, this lack of execution of software during analysis has its limitations. It cannot check for such problems as deadlocks, data race conditions and memory leaks.

Another example of a tool used for static analysis from the open source domain is the XRadar tool, which includes both the static and dynamic execution analysis. An example screenshot is presented in Fig. 5.35.

If the software development is done in the Eclipse environment (www.eclipse.org) then there are over 1000 plug-ins which provide the ability to statically analyze the software code. Many of these plug-ins implement the MISRA standard checks.

5.8.3   Testing

Testing is also a very well-known technique which should be mentioned here. However, we've already discussed it in Chap. 3.

5.9   Further Reading

Readers who are interested in more hardware-software integration and programming for automotive systems can study the book by Schauffele and Zurawka [SZ05]. They describe in more detail the concepts used in the detailed design of automotive software, such as timing analysis and hardware-oriented programming.

A good read for software engineers who move into the field of automotive software design is the book chapter by Saltzmann and Stauner [SS04], who describe the specifics of automotive software development compared to non-automotive development.

For modelling in Simulink the best resource is the website of Matlab with its numerous tutorials—www.matlab.com. In order to strengthen one's understanding of the process of translating the physical world to the Simulink models, we recommend the tutorial from https://classes.soe.ucsc.edu/cmpe242/Fall10/simulink. pdf.

More advanced readers who are seeking methods for optimizing Simulink models should look at the article by Han et al. [HNZ$^+$13], who focus on that topic discussing areas such as, for example, hydraulic servo mechanism. Another good read in this direction, about detection of model smells, is the paper by Gerlitz et al. [GTD15].

The MISRA standard is a well-known one, but it has been developed taking into consideration NASA's 10 rules of safety-critical programming [Hol06]. The rationale and empirical evidence of using smaller sets of language constructs in safety-critical systems can be found in the article by Hatton [Hat04].

Readers who are interested in a more detailed description of programming languages and principles used in safety-critical programming can refer to Fowler's compendium [Fow09] or to the classical position by Storey [Sto96]. We also recommend our previous work on the evolution of complexity of automotive software [ASM$^+$14] and its impact on reliability [RSM$^+$13].

Using formal methods in the design of automotive software has been shown to be efficient to validate product configurations as the number of all potential variants is large, while the number of allowed variants is much smaller. Sinz et al. have shown one such application [SKK03]. Another area is the integration of software as shown by Jersak et al. [JRE$^+$03].

As using formal methods in general is rather costly, researchers constantly seek new ways of decreasing cost, for example, by searching for lightweight methods, such as the one advocated by Jackson [Jac01].

For readers interested in using and customizing UML for the purpose of detailed design of automotive software I recommend taking a look at our previous work on the impact of different customization mechanisms on the quality of models [SW06] and the process of realizing MDA in industry [SKW04, KS02] and the problems of inconsistent designs [KS03].

Finally, readers interested in the quality of automotive software may find it interesting to study defect classification schemes, where the attributes of faults encountered in automotive software are described in more detail [MST12].

5.10 Summary

Since automotive software consists of multiple domains and multiple types of computers, detailed design of it is based on many different paradigms, which we briefly introduced in this chapter.

In this chapter we have explored ways in which software designers work with detailed design of automotive software. We have focused on model-based development using Simulink, which is the most common design tool and method for the automotive software.

We have also introduced the principles of programming of safety-critical systems, which are based on NASA's principles and the MISRA standard. In short, these principles postulate the need to use simple programming constructs which allow us to verify the validity of the program before its execution and minimize the risk of unexpected behaviour of the software.

In this chapter we have also looked at the GENIVI architecture of infotainment systems, which is one of the interesting areas in automotive software. Finally, towards the end of the chapter we looked at a number of different techniques for verifying automotive software, such as static analysis and formal verification.

References

A⁺08.   Motor Industry Software Reliability Association et al. *MISRA-C: 2004: guidelines for the use of the C language in critical systems*. MIRA, 2008.

All09.   GENIVI Alliance. Genivi, 2009.

All14.   GENIVI Alliance. Bmw case study, 2014.

All15.   GENIVI Alliance. Reference architecture, 2015.

ASM⁺14.   Vard Antinyan, Miroslaw Staron, Wilhelm Meding, Per Österström, Erik Wikstrom, Johan Wranker, Anders Henriksson, and Jörgen Hansson. Identifying risky areas of software code in agile/lean software development: An industrial experience report. In *Software Maintenance, Reengineering and Reverse Engineering (CSMR-WCRE), 2014 Software Evolution Week-IEEE Conference on*, pages 154–163. IEEE, 2014.

BV01.   Guillaume Brat and Willem Visser. Combining static analysis and model checking for software analysis. In *Automated Software Engineering, 2001.(ASE 2001). Proceedings. 16th Annual International Conference on*, pages 262–269. IEEE, 2001.

EM04.   Dawson Engler and Madanlal Musuvathi. Static analysis versus software model checking for bug finding. In *International Workshop on Verification, Model Checking, and Abstract Interpretation*, pages 191–210. Springer, 2004.

Fow09.   Kim Fowler. *Mission-critical and safety-critical systems handbook: Design and development for embedded applications*. Newnes, 2009.

GTD15.   Thomas Gerlitz, Quang Minh Tran, and Christian Dziobek. Detection and handling of model smells for MATLAB/Simulink Models. In *Proceedings of the International Workshop on Modelling in Automotive Software Engineering. CEUR*, 2015.

Hat04.   Les Hatton. Safer language subsets: An overview and a case history, MISRA C. *Information and Software Technology*, 46(7):465–472, 2004.

HNZ⁺13.   Gang Han, Marco Di Natale, Haibo Zeng, Xue Liu, and Wenhua Dou. Optimizing the implementation of real-time simulink models onto distributed automotive architectures. *Journal of Systems Architecture*, 59(10, Part D):1115–1127, 2013.

Hol06.   Gerard J Holzmann. The power of 10: rules for developing safety-critical code. *Computer*, 39(6):95–99, 2006.

ISO16.   ISO/IEC. ISO/IEC 25000 - Systems and software engineering - Systems and software Quality Requirements and Evaluation (SQuaRE). Technical report, 2016.

Jac01.   Daniel Jackson. Lightweight formal methods. In *International Symposium of Formal Methods Europe*, pages 1–1. Springer, 2001.

JRE⁺03.   Marek Jersak, Kai Richter, Rolf Ernst, J-C Braam, Zheng-Yu Jiang, and Fabian Wolf. Formal methods for integration of automotive software. In *Design, Automation and Test in Europe Conference and Exhibition, 2003*, pages 45–50. IEEE, 2003.

Kni02.   John C Knight. Safety critical systems: Challenges and directions. In *Software Engineering, 2002. ICSE 2002. Proceedings of the 24th International Conference on*, pages 547–550. IEEE, 2002.

KS02.   Ludwik Kuzniarz and Miroslaw Staron. On practical usage of stereotypes in UML-based software development. *the Proceedings of Forum on Design and Specification Languages, Marseille*, 2002.

KS03.   Ludwik Kuzniarz and Miroslaw Staron. Inconsistencies in student designs. In *the Proceedings of The 2nd Workshop on Consistency Problems in UML-based Software Development, San Francisco, CA*, pages 9–18, 2003.

MST12.   Niklas Mellegård, Miroslaw Staron, and Fredrik Törner. A light-weight software defect classification scheme for embedded automotive software and its initial evaluation. *Proceedings of the ISSRE 2012*, 2012.

RSM⁺13.   Rakesh Rana, Miroslaw Staron, Niklas Mellegård, Christian Berger, Jörgen Hansson, Martin Nilsson, and Fredrik Törner. Evaluation of standard reliability growth models in the context of automotive software systems. In *Product-Focused Software Process Improvement*, pages 324–329. Springer, 2013.

SKK03.   Carsten Sinz, Andreas Kaiser, and Wolfgang Küchlin.   Formal methods for the validation of automotive product configuration data. *AI EDAM: Artificial Intelligence for Engineering Design, Analysis and Manufacturing*, 17(01):75–97, 2003.

SKW04.   Miroslaw Staron, Ludwik Kuzniarz, and Ludwik Wallin. Case study on a process of industrial MDA realization: Determinants of effectiveness. *Nordic Journal of Computing*, 11(3):254–278, 2004.

SS04.   Christian Salzmann and Thomas Stauner. *Automotive Software Engineering*, pages 333–347. Springer US, Boston, MA, 2004.

Sta93.   ISO Standard. ISO 11898, 1993. *Road vehicles–interchange of digital information–Controller Area Network (CAN) for high-speed communication*, 1993.

Sto96.   Neil R Storey. *Safety critical computer systems*. Addison-Wesley Longman Publishing Co., Inc., 1996.

SW06.   Miroslaw Staron and Claes Wohlin. An industrial case study on the choice between language customization mechanisms. In *Product-Focused Software Process Improvement*, pages 177–191. Springer, 2006.

SZ05.   Jörg Schäuffele and Thomas Zurawka. *Automotive software engineering – Principles, processes, methods and tools*. 2005.

Chapter 6
Evaluation of Automotive Software Architectures

Abstract In this chapter we introduce methods for assessing the quality of software architectures and we discuss one of the techniques—ATAM. We discuss the non-functional properties of automotive software and we review the methods used to assess such properties as dependability, robustness and reliability. We follow the ISO/IEC 25000 series of standards when discussing these properties. In this chapter we also address the challenges related to the integration of hardware and software and the impact of this integration. We review differences with stand-alone desktop applications and discuss examples of these differences. Towards the end of the chapter we discuss the need to measure these properties and introduce the need for software measurement.

6.1 Introduction

Having the architecture in place, as we discussed in Chap. 2, is a process which requires a number of steps and revisions of the architecture. As the evolution of the architecture is a natural step, it is often guided by some principles. In this chapter we look into aspects which drive the evolution of the architectures—non-functional requirements and architecture evaluation methods.

During this process the architects take a number of decisions about their architecture—starting from the basic one on what style should be used in which part of the architecture and ending in the one on the distribution of signals over the car's communication buses. All of these evaluations lead to a better or worse architecture and in this chapter we focus on the question that each software architect confronts—*How good is my architecture?*

Although the question is rather straightforward, the answer to it is rather complicated, because the answer to it depends on a number of factors. The major complication is related to the need to balance all of these factors. For example, the performance of the software needs to be balanced with the cost of the system, the extensibility needs to be balanced with the reliability and performance, etc. Since the size of the software system is often large the question whether the architecture is optimal, or even good enough, requires an organized way of evaluating the architecture.

In Chap. 3 we discussed the notion of a requirement as a customer demand on the functionality of the software and the need for the fulfillment of certain quality

© Springer International Publishing AG 2017
M. Staron, *Automotive Software Architectures*,
DOI 10.1007/978-3-319-58610-6_6

attributes. In this chapter we dive deeper into the question—*What quality attributes are important for the automotive software architectures?* and *How do we evaluate that an architecture fulfills these requirements?*

To answer the first question we review the newest software engineering standard in the area of product quality—ISO/IEC 25023 (Software Quality Requirements and Evaluation—Product Quality, [ISO16b]). We look into the construction of the standard and focus on how software quality is described in this standard, with the particular focus on product quality.

To answer the second question about the evaluation of architectures, we look into one of the techniques for evaluating quality of software architectures—Architecture Trade-off Analysis Method (ATAM), which is one of the many techniques for assessing quality of software architectures.

So, let us dive deeper into the question of what software quality is and how it is defined in modern software engineering standards.

6.2 ISO/IEC 25000 Quality Properties

One of the main standards in the area of software quality is the ISO/IEC 25000 series of standards—Software Quality Requirements and Evaluation (SQuaRE) [ISO16a]. The standard is an extension of the old standard in the same area—ISO/IEC 9126 [OC01]. Historically, the view of the software quality concept in ISO/IEC 9126 was divided into a number of sub-areas such as reliability or correctness. This view was found to be too restrictive as the quality needs to be related to the context of the product—its requirements, operating environment and measurement. Therefore, the new ISO/IEC 25000 series of standards is more extensive and has a modular architecture with a clear relation to other standards. An overview of the main quality attributes, grouped into quality characteristics, is presented in Fig. 6.1. The dotted line shows a characteristic which is not part of the ISO/IEC 25000 series, but another standard—ISO/IEC 26262 (Road Vehicles–Functional Safety).

These quality characteristics describe various aspects of software quality, such as whether it fulfills the functions described by the requirements correctly (functionality) and whether it is easy to maintain (maintainability). However, for safety-critical systems like the software system of a car, the most important part of the quality model is actually the reliability part, which defines the reliability of a software system, such as *Degree to which a system, product or component performs specified functions under specified conditions for a specified period of time*, [ISO16b].

6.2.1 Reliability

Reliability of a software system in common understanding is the ability of the system to work according to the specification during a period of time [RSB+13].

Fig. 6.1 ISO/IEC 25000 quality attributes

This characteristic is important as car's computer system, including software, has to be in operation for years after its manufacturing. The ability to "reset" the car's computer system is very limited as it needs to operate constantly, controlling the powertrain, brakes, and safety mechanisms.

Reliability is a generic quality characteristics and contains four sub-characteristics as shown in Fig. 6.2—maturity, availability, recoverability and fault tolerance.

Maturity is defined as *degree to which a system, product or component meets needs for reliability under normal operation*. The concept defines how the software operates over time, i.e. how many failures the software has over time, which is often shown as a curve of the number of defects over time; see Fig. 6.2 from [RSM+13] and [RSB+16].

The figure shows that the number of faults discovered during the design and operation of the software system can have different shapes depending on the type of development, type of the functionality being developed and the time of the lifecycle of the software. The type of development (discussed in Chap. 3) determines how and when the software is tested and the testing determines the type of faults that are discovered—e.g. the late testing phases often uncovers more severe defects, while the early testing phases can isolate simpler defects that can be fixed easily. Flattening

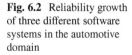

Fig. 6.2 Reliability growth of three different software systems in the automotive domain

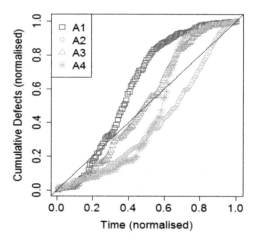

of the curve towards the end of the development shows that the maturity of the system is higher as the number of defects found gets lower—the software is ready for its release and deployment.

Another sub-characteristic of reliability is the availability of the system, which is defined as *degree to which a system, product or component is operational and accessible when required for use*. The common sense of this definition is the ability of the system to be used when needed, which can be seen as a momentary property. High availability systems do not need to be available over time, all the time, but they need to be available when needed. This means that these systems can be restarted often and the property of "downtime" is not as important as for fault-tolerant systems which should be available all the time (e.g. 99.999% of the time, which is ca. 4 min of downtime per year).

Recoverability is defined as *Degree to which, in the event of an interruption or a failure, a product or system can recover the data directly affected and re-establish the desired state of the system*. This quality property is often quoted in the research on self-* systems (e.g. self-healing, self-adaptive, self-managing) where the software itself can adjust its structure in order to recover from failure. In the automotive domain, however, this is still in the research phase as the mechanisms of self-* often should be formally proven that the transition between states is safe. The only exception is the ability of the system to restart itself, which has been used as a "last resort" mechanism for tackling failures.

Fault tolerance is defined as *degree to which a system, product or component operates as intended despite the presence of hardware or software faults*. This property is very important as the car's software consists of hundreds of software components distributed over tens of ECUs communicating over a few buses—something is bound to go wrong in this configuration. Therefore we discuss this property separately in the next section.

Fig. 6.3 Engine check control light indicating reduced performance of the powertrain, Volvo XC70

6.2.2 Fault Tolerance

Fault tolerance, or robustness is a concept of *the degree to which a computer system can operate in the presence of errors* [SM16]. Robustness is important as the software system of a car needs to operate, sometimes with reduced functionality, even if there are problems (or errors) during runtime.

A common manifestation of the robustness of the car is the ability to operate with reduced functionality when the diagnostics system indicates a problem with, for example, the powerline. In many modern cars the diagnostics system can detect problems with the exhaust system and reduce the power of the engine (degradation of the functionality), but still enable the operation of the car. The driver is only notified by a control lamp on the instrument panel as in Fig. 6.3.

As the figure shows, the software system (the diagnostics) has detected the problem and has taken action to allow the driver to continue the journey—which shows high robustness to failures.

6.2.3 Mechanisms to Achieve Reliability and Fault Tolerance

The traditional ways of achieving fault tolerance are often found on the lower levels of system design—hardware level. The ECUs used in the computer system can rely on hardware redundancy and fail-safe takeover mechanisms in order to ensure the operation of the system in the presence of faulty component. However, this approach is often non-feasible in the car's software as the electrical system of the car cannot be duplicated and hardware redundancy is not possible. Instead, the designers of the software systems usually rely on substituting data from different sensors in order to obtain the same (or similar) information once one of the components fails.

One of the main mechanisms used in modern software is the mechanism of *graceful degradation*. Shelton and Koopman [SK03] define graceful degradation as *a measure of the system's ability to provide its specified functional and non-functional capabilities*. They show that a system that has all of its components functioning properly has maximum utility and "losing" one or more components leads to reduced functionality. They claim that "a system degrades gracefully if individual component failures reduce system utility proportionally to the severity of aggregate failures." For the architecture, this means that the following decisions need to be prioritized:

- No single point of failure—this means that no component should be exclusively dependent on the operation of another component. Service-oriented architectures and middleware architectures often do not have a single point of failure.
- Diagnosing the problems—the diagnostics of the car should be able to detect malfunctioning of the components, so mechanisms like heartbeat synchronization should be implemented. The layered architectures support the diagnostics functionality as they allow us to build two separate hierarchies—one for handling functionality and one for monitoring it.
- Timeouts instead of deadlocks—when waiting for data from another component, the component under operation should be able to abort its operation after a period of time (timeout) and signal to the diagnostics that there was a problem in the communication. Service-oriented architectures have built-in mechanisms for monitoring timeouts.

Prioritizing such decisions should lead to an architecture where a single failure in a component leaves the entire system operational and signals the need for manual intervention (e.g. workshop visit to replace a faulty component).

A design principle to achieve fault-tolerant software is to use programming mechanisms which reduce the risk of both design and runtime errors, such as:

- using static variables when programming—using static variables rather than variables allocated dynamically on the heap allows taking advantage of atomic write/read operations; when addressing a memory dynamically on the heap the read/write operation requires at least two steps (read the memory address, write/read to the address), which can pose threats when using multithreaded programs or interrupts.
- using safety bits for communication—any type of communication should include the so-called safety bits and checksums in order to prevent operation of software components based on faulty inputs and thus failure propagation.

The automotive industry has adopted the MISRA-C standard, where the details of the design of computer programs in C programming language [A+08], which has been discussed in more detail in the previous chapter.

However, since the architecture of the software is an artifact that is abstract and cannot be tested, the evaluation of the architecture needs to be done based on its description as a model and often manually.

6.3 Architecture Evaluation Methods

In our discussion of the quality of the system we highlighted the need to balance different quality characteristics against each other. This balancing needs to be evaluated and therefore we look into an example software architecture evaluation technique.

The goals behind evaluating architectures can differ from case to case, from the general understanding of the architectural principles to the exploration of specific risks related to software architectures. Let us explore what kinds of architecture analysis methods are the most popular today and why.

Techniques used for analysis of architectures, as surveyed by Olumofin [OM05]:

1. Failure Modes and Effects Analysis (FMEA)—a method to analyze software designs (including the architecture) from the perspective of risk of failures of the system. This method is one of the most generic ones and can come either in fully qualitative form (based on expert analysis) or as a combination of qualitative expert analysis and quantitative failure analysis using mathematical formulas for failure modelling.
2. Architecture Trade-off Analysis Method (ATAM)—a method to evaluate software architectures from the perspective of the quality goals of the system. ATAM, based on expert-based reviews of the architecture from the perspective of scenarios (more about it later in this chapter).
3. Software Architecture Analysis Method (SAAM)—a method which is seen as a precursor to ATAM is based on the evaluation of software architectures from the perspective of different types of modifiability, portability and extendability. This method has multiple variations, such as: SAAM Founded on Complex Scenarios (SAAMCS), Extending SAAM by Integration in the Domain (ESAAMI) and Software Architecture Analysis Method for Evolution and Reusability (SAAMER).
4. Architecture Level Modifiability Analysis (ALMA)—a method for evaluating the ability of the software architecture to withstand continuous modifications, [BLBvV04].

The above evaluation methods constitutes an important method portfolio for software architects who need to make judgements about the architecture of the system before the system is actually implemented. It seems like a straightforward task, but in reality it requires skills and experience to be performed correctly.

An example of the need for skills and experiences is the evaluation of the performance of the system before it is implemented. When designing the software system

Fig. 6.4 Parking assistance camera showing the view behind the car while backing up, Volvo XC70

in cars the performance of the communication channels is often a bottleneck—the bandwidth of the CAN bus is usually limited. Therefore adding new, bandwidth-greedy components and functions requires analysis of both the scenario of using the function in question and the entire system. A simple case is the function of providing a camera video feed from the back of the car when backing-up—used in the majority of premium segment cars today. Figure 6.4 shows this function on the instrument panel.

When adding the camera to the electrical system the amount of data transmitted from the back of the car to the front of the car increases dramatically (depending on the resolution of the camera, it could be up to 1 Mbit/s). Since the data is to be transmitted in real time the communication bus must constantly prioritize between the video feed data and the signals from such sensors as parking assist sensors.

In this scenario the architects need to answer the question—will it be possible to add the camera component to the electrical system without jeopardizing such safety critical functions as park assist?

6.4   ATAM

ATAM has been designed as a response to the need of the American Department of Defense in the 1990s to be able to evaluate the quality of software systems in their early development stage (i.e. before the system is implemented). The origins of ATAM are at the Software Engineering Institute, in the publication of Kazman et al. [KKB+98]. The ATAM method, which can be used to answer this question is based on [KKC00]:

The Architecture Tradeoff Analysis Method (ATAM) is a method for evaluating software architectures relative to quality attribute goals. ATAM evaluations expose architectural risks that potentially inhibit the achievement of an organization's business goals. The ATAM gets its name because it not only reveals how well an architecture satisfies particular quality goals, but it also provides insight into how those quality goals interact with each other and how they trade off against each other.

As stressed in the above definition, the method relates the system to its quality, i.e. non-functional requirements on its performance, availability, reliability (fault tolerance) and other quality characteristics of ISO/IEC 25000 (or any other quality model).

6.4.1 Steps of ATAM

ATAM is a stepwise method which is similar to reading techniques used in software inspections (e.g. perspective-based reading [LD97] or checklist-based reading [TRW03]). The steps are as follows (after [KKC00]).

Step 1: Present ATAM. In this step the architecture team presents the ATAM method to the stakeholders (architects, designers, testers and product managers). The presentation should explain the principles of the evaluation, evaluation scenarios and its goal (e.g. which quality characteristics should be prioritized).

Step 2: Present business drivers. After presenting the purpose of the evaluation, the purpose of the business behind this architecture is presented. Topics covered in this step should include: (i) the main functions of the system (e.g. new car functions), (ii) the business drivers behind these functions and their optionality (e.g. which functions are to be included in all models and which should be optional), business case behind the architecture and its main principles (e.g. performance over extendability, maintainability over cost).

Step 3: Present architecture. The architecture should be presented in a sufficient level of detail to make the evaluation. The designers of the ATAM method do not propose a specific level of detail, but it is customary that the architects guide the reading of the architecture model—show where to start and where to stop reading the architecture model.

Step 4: Identify architectural approaches. In this step the architects introduce the architectural styles to the analysis team and present the high-level rationale behind these approaches.

Step 5: Generate quality attribute utility tree. In this step, the evaluation team constructs the system utility measure tree by combining the relevant quality factors, specified with scenarios, stimuli and responses.

Step 6: Analyze architectural approaches. This is the actual evaluation step where the evaluation team explores the architecture by studying the prioritized scenarios from step 5 and architectural approaches which address these scenarios and their corresponding quality characteristics. This step results in identifying architectural risks, sensitivity points, and tradeoff points.

Step 7:   Brainstorm and prioritize scenarios. After the initial analysis of the architectural approaches is done, there is a lot of scenarios and sensitivity points elicited from the evaluation team. Therefore they need to be prioritized to guide the further analysis of the architecture. The 100 dollar technique, planning game and analytical-hierarchy-process are useful prioritization techniques at this stage.

Step 8:   Analyze architectural approaches. In this step the team reiterates the analysis from step 6 with a focus on the highly prioritized scenarios from step 7. The result is again the list of risks, sensitivity points and trade-off points.

Step 9:   Present results. After the analysis the team compiles and presents a report about the found risks, sensitivity points, non-risks and tradeoffs in the architecture.

The results of the analysis can only be as good as the input to the analysis, i.e. the quality of the architecture documentation (its completeness and correctness), the quality of the scenarios, the templates used in the analysis and the experience of the evaluation team.

6.4.2   Scenarios Used in ATAM in Automotive

ATAM is an extensible method which allows us to identify scenarios by the evaluation team, which is strongly encouraged. In this chapter we present a set of inspirational scenarios to guide the evaluation team. Our example set is based on the example set of scenarios presented by Bass et al. [BM$^+$01] and in this chapter we present a set of scenarios important for the evaluation of automotive software. We present them in generic terms and in compact textual format. We group them according to quality characteristics, following the approach presented by Bass et al.

6.4.2.1   Modifiability

We start with the set of scenarios which date back to the origins of ATAM and address one of the main challenges for the work of the software architects—How extendable and modifiable is our architectural design?

It is worth noting that some of the scenarios impact the design (or the internal quality) of the product and some impact the external quality. The modifiability scenarios impact the internal quality of the product.

Scenario 1:   A request arrives to change the functionality of the system. The change can be to add new functionality, to modify existing functionality, or to delete functionality [BM$^+$01].

Scenario 2:   A request arrives to change one of the components (e.g. because of a technology shift); the scenario needs to consider the change propagation to the other components.

Scenario 3:   Customer wants different systems with different capabilities but using the same software and therefore advanced variability has to be built into the system [BM+01].

Scenario 4:   New emission laws: the constantly changing environmental laws require adaptation of the system to decrease its environmental impact [BM+01].

Scenario 5:   Simpler engine models: Replace the engine models in the software with simple heuristics for the low-cost market [BM+01].

Scenario 6:   An additional ECU is added to the vehicle's network and causes new messages to be sent through the existing network. In the scenario we need to understand how the new messages impact the performance of the entire system.

Scenario 7:   An existing ECU after the update adds a new message type: same messages but with additional fields that we are currently not set up to handle (based on [BM+01]).

Scenario 8:   A new AUTOSAR version is adopted and requires update of the base software. We need to understand the impact of the new version in terms of the number of required modifications to the existing components.

Scenario 9:   Reduce memory: During development of an engine control, the customer demands we reduce costs by downsizing the flash-ROM on chip (adapted from [BM+01]). We need to understand what the impact of this reduction is on the system performance.

Scenario 10:   Continuous actuator: Changing two-point (on/off) actuators to continuous actuators within 1 month (e.g., for the EGR or purge control valve). We need to understand the impact of this change on the behavior of our models [BM+01].

Scenario 11:   Multiple engine types in one car need to coexist: hybrid engine. We need to understand how to adapt the electrical system and isolate the safety-critical functions from the non-safety-critical ones.

6.4.2.2 Availability and Reliability

Availability and reliability scenarios impact the external quality of the product—allow us to reason about the potential defects which come from unfulfilled performance requirements (non-functional requirements).

Scenario 12:   A failure occurs and the system notifies the user; the system may continue to perform in a degraded manner. What graceful degradation mechanisms exist? (based on [BM+01]).

Scenario 13:   Detect software errors existing in third-party or COTS software integrated into the system to perform safety analysis [BM+01].

6.4.2.3   Performance

Performance scenarios also impact the external quality of the product and allow us to reason about the ability of the system to fulfill performance requirements.

Scenario 14:   Start the car and have the system active in 5 s (adapted from [BM$^+$01]).
Scenario 15:   An event is initiated with resource demands specified and the event must be completed within a given time interval [BM$^+$01].
Scenario 16:   Using all sensors at the same time creates congestion and this causes loss of safety-critical signals.

6.4.2.4   Developing Custom Scenarios

It is natural that during an ATAM assessment the assessment group combines standard scenarios with custom ones. The literature about ATAM encourages us to create custom scenarios and use them in the evaluations, and therefore a few key points emerge which can help the development of scenarios.

Scenarios should be relevant to both the quality model's chosen/prioritized quality attributes and the business model of the company. It is important that the evaluation of the architecture be done in order to ensure that it fulfills the boundaries of product development. The BAPO model (Business Architecture Process and Organization, [LSR07]) from the evaluation of product lines can be used to make the link.

The criteria applied for the scenarios should be clear to the assessment team and the organization. It is important that all stakeholders understand what "good", "wrong", "insufficient", and "enough" mean in the evaluation situation. It is all too easy to get stuck in a detailed discussion of mechanisms used in the evaluation without the good support of measures or checklists.

When defining custom scenarios we can get help of the table with the elements presented in Fig. 6.5.

6.4.3   Templates Used in the ATAM Evaluation

The first template which is needed in the ATAM evaluation is the template to specify the scenarios. An example scenario template is presented in Fig. 6.6.

One of the templates, needed after the ATAM evaluation is completed, is the risk description template, which should be included in the results and their presentation. An example template is presented in Fig. 6.7.

Another part of the results from ATAM is the set of sensitivity points which have been found in the architecture. A sensitivity point is defined by the Software

Aspect	Value
Source	The description of which architectural element initiates the scenario.
Stimulus	The stimulus signal or component of the scenario.
Artifact	Architectural elements which are affected by the scenario.
Environment	Description of the environment when this stimulus appears.
Response	Description of the expected outcome observed after the received stimulus.
Measure	Quantifiable measures that could help if the scenario is successful.

Fig. 6.5 Template for defining custom scenarios

Scenario ID	Unique ID of the scenario to identify it, later on used to link the scenario to the quality characteristics, and requirements
Stimulus	The stimulus in the scenario, i.e. what kind of event or activity of interest in the scenario. For example: Adding a new rear-view camera to the main CAN bus.
Response	The outcome of interest in the scenario. For example: Causes the congestion of signals on the bus and loss of safety critical signals from the parking assist sensors.
Requirement	The link of the scenario to the requirement(s) of the architecture, its performance or other non-functional characteristics.
Quality characteristics	The link of the scenario to one of the quality characteristics, e.g. modifiability, safety.
Textual version (optional)	Combining the stimulus and response into one sentence. For example: *"Adding a new rear-view camera to the main CAN bus can cause the congestion of signals on the bus and thus loss of safety-critical signals."*

Fig. 6.6 Template for the description of a scenario in ATAM

Engineering Institute as

"*a property of one or more components (and/or component relationships) that is critical for achieving a particular quality attribute response. Sensitivity points are places in a specific architecture to which a specific response measure is particularly sensitive (that is, a little change is likely to have a large effect). Unlike tactics, sensitivity points are properties of a specific system.*"

A tradeoff template is presented in Fig. 6.8.

Risk ID	Unique ID for the identification of the risk
Description	Detailed description of the risk, including the source of the risk.
Source / Sensitivity point	The description of the source of the risk. This field should include the reference to the element of the architecture which is the source of the risk in question. A detailed reference is important as it is needed for the assessment of the safety of the software system.
Impact	The description of the impact of the risk on the scenario, the quality characteristics of the system and ultimately the user of the system. For the risks related to the safety-critical functions of the system (e.g. when the ASIL level D is assigned to the source component), this impact should be related to the appropriate ASIL level requirements.
Severity	Severity of the risk, usually on the scale 1-5 from the least severe to critical.
Probability	The probability that this risk will manifest itself in the runtime system, usually on the scale 1-5 from the very unlikely to certain.

Fig. 6.7 Template for the description of risks found in ATAM

Tradeoff ID	The ID of the tradeoff.
Quality characteristic 1	The first characteristic which is taking part of the trade-off.
Quality characteristic 2	The second characteristic which is taking part of the trade-off.
Sensitivity point	The sensitivity point in the software architecture where the trade-off decision takes place.
Tradeoff description	The description of the rationale and reasoning behind the trade-off. Here, the evaluation team should describe why this is trade-off identified and how the changes in the architecture to address one of the quality characteristics affect the other one.

Fig. 6.8 Template for the description of trade-offs identified after the ATAM analysis

6.5   Example of Applying ATAM

Now that we have reviewed the elements of ATAM and its process, let us illustrate ATAM analysis using the example of placing the functionality related to a rear-view camera on the back bumper of the car. As we have just introduced ATAM in this chapter, let us start with the introduction of the business drivers.

6.5.1   Presentation of Business Drivers

The major business driver in this architecture is achieving a high degree of safety.

6.5.2   Presentation of the Architecture

First, let us present the function architecture of the car in Fig. 6.9.

Since we focus on camera functionality, we only include the major functions from the domains of active safety and infotainment. The functions presented in the figure represent the basic functions of braking and ABS in the active safety domain and the displaying of information on screens (both the main screen and the head-up display HUD).

Let us now introduce the simplistic architecture of the car's electrical system—i.e. the physical view of the architecture. The physical view is presented in Fig. 6.10.

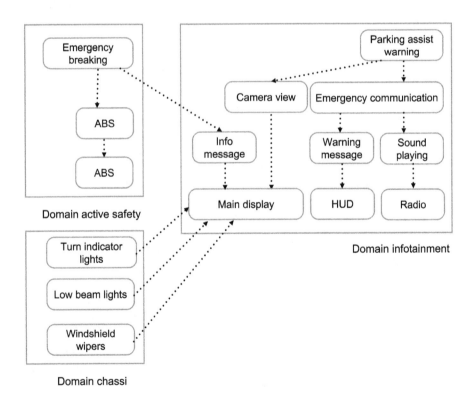

Fig. 6.9 Function dependencies in the architecture in our example

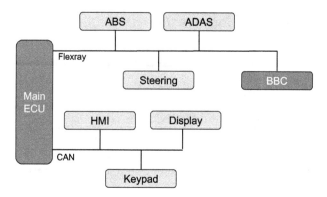

Fig. 6.10 Physical view of the architecture in our example

In the example architecture we have two buses:

- CAN bus: connecting the ECUs related to the infotainment domain.
- Flexray bus: connecting the ECUs related to the safety domain and the chassi domain

We can also see the following ECUs :

- Main ECU: the main computer of the car, controlling the configuration of the car, initialization of the electronics and diagnostics of the entire system. The main ECU has the most powerful computing unit in the car, with the largest memory (in our example).
- ABS (Anti-locking Brake System): the control unit responsible for the braking system and the related functionality; it is a highly safety-critical unit, with only the highest safety integrity level software.
- ADAS (Advanced Driver Assistance and Support): the control unit responsible for higher-level decisions regarding active safety, such as collision avoidance by braking, emergency braking and skid prevention; it is also responsible for such functions as parking assistance.
- Steering: the control unit responsible for the steering functionality such as the electrical servo; it is also the controller of parts of the functions or parking assistant.
- BBC (Back Body Controller): the unit responsible for controlling non-safety critical functions related to the back of the car, such as adjusting of anti-dim lights, turning on and off of blinkers (back), and electrical opening of the trunk.

In the logical view of the architecture we focus on showing the main components used in the display of information and its processing from the camera unit, as we need them to perform the architecture analysis. Now let us introduce the logical architecture of the system in Fig. 6.11.

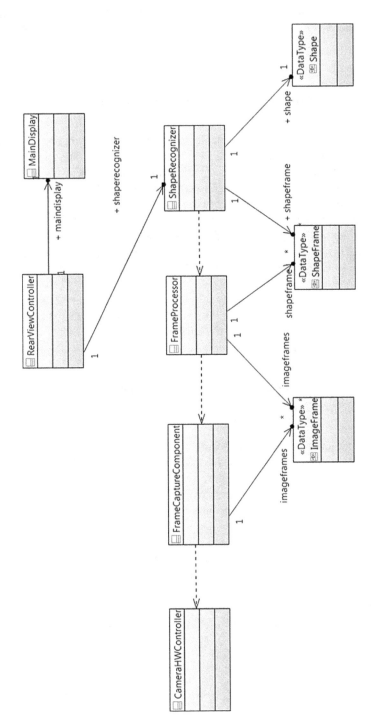

Fig. 6.11 Logical view of the architecture in our example

And finally let us show the potential deployment alternative of the architecture, where the majority of the processing takes place in the BBC node—as we can see in Fig. 6.12.

6.5.3   Identification of Architectural Approaches

In this example let us focus on the deployment of software components on the target ECUs. We also say that the physical architecture (hardware) does not change and therefore we analyze the software aspects of the car's electrical system. As an alternative approach let us consider deploying all the processes on the main ECU instead of dividing the components between the Main ECU and the BBC. This results in the deployment as shown in Fig. 6.13. The dominant architectural style is pipes-and-filters as the processing of images is the main functionality here. The car's electrical system should support the advanced mechanisms of active safety (i.e. controlled by software) and should ensure that none of the mechanisms interfere with another one, jeopardizing safety.

In our subsequent considerations we look into these two alternatives and decide which one should be chosen to support the desired quality goals—i.e. what decision the architect should take given his quality attribute tree.

6.5.4   Generation of Quality Attribute Tree and Scenario Identification

In this example let us consider two scenarios which complement each other. We could naturally generate many more for each of the quality attributes presented earlier in this chapter, but we focus on the safety attribute—a scenario where there is congestion on the CAN bus when reverse driving and using a camera, and a scenario where we overload the main ECU when the video feed computations can interfere with other functions such as the operation of windshield wipers and low beam lights. We can use the scenario description template to outline the scenario in Fig. 6.14.

Let us also fully describe the first scenario as presented in Fig. 6.15.

In this scenario we are interested in the safety aspect of the reverse camera. We need to understand what kind of implications the video feed data transfer has on the capacity of the CAN bus which connects the BBC computer with the main ECU. We therefore need to consider both alternative architectural decisions—deployment of the video processing functionality on the BBC and the main ECU. We assume

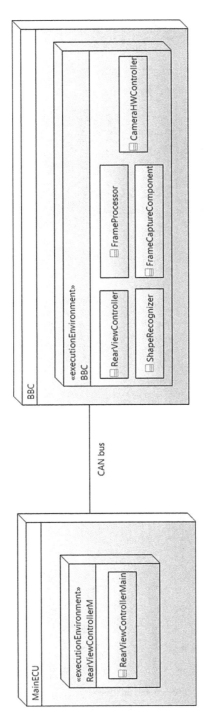

Fig. 6.12 The first deployment alternative in our example

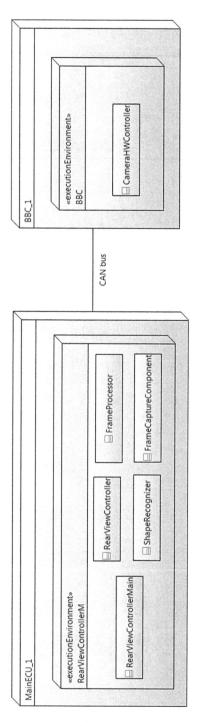

Fig. 6.13 The second deployment alternative in our example

Aspect	Value
Source	Rear camera.
Stimulus	Camera feed.
Artifact	Main ECU, BBC ECU, CAN Bus.
Environment	Car in reverse driving.
Response	Process video data and show it on the display.
Measure	Video displayed in real time and no loss of safety signals from the parking sensors.

Fig. 6.14 Scenario described with its stimulus, response, environment and measure

Scenario ID	SC1: Congestion on the bus during reverse driving prevents safety-critical signals from reaching their destination.
Stimulus	The scenario is that during the reverse driving (backing up) of the car the video feed from the rear camera uses too much of the capacity and the communication bus is not able to relay (send) signals from the parking sensors. The main question to evaluate in this scenario is what kind of software deployment has the lowest influence on the safety of the car's software?
Response	• Analysis of the potential congestion for two architecture deployments. • List of constraints on the functionality for each of the solutions.
Requirement	"The architecture should allow the safety critical signals to be sent/received at any given point of time."
Quality characteristics	Safety: in this scenario we need to know that the particular architecture of the software does not cause congestions on buses and potential loss of signals.
Textual version (optional)	*When reversing the car, the video feed from the camera can reduce the ability of the parking sensors to send signals to the main ECU and therefore do not warn the driver about the potential collision.*

Fig. 6.15 Scenario of congestion on the communication bus

that none of the deployments result in adding new hardware and therefore do not influence the performance of the electrical system as a whole.[1]

We also can identify a scenario which is complementary to this one—see Fig. 6.16.

---

[1]This assumption simplifies the analysis as we do not need to consider the physical architecture, but can focus only on the logical and deployment views of the architecture.

Scenario ID	SC2: Overloading of the main processor during heavy weather conditions reduces the quality of the video feed.
Stimulus	The scenario is that during the heavy rain/snow condition where the main ECU is responsible for steering the windshield wipers, operating the lights and processing the video feed, the processing power of the ECU might not be enough to cope with all calculations The main question to evaluate in this scenario is, what kind of software deployment has the lowest influence on the performance of the car's software?
Response	• Analysis of the potential processing power for two architecture deployments. • List of constraints on the functionality for each of the solutions.
Requirement	"The car should provide the video feed from the rear-view camera during reverse driving in all weather conditions."
Quality characteristics	Performance: in this scenario we need to know that the particular architecture of the software does not cause overload of the computers and thus reduce the quality of the video feed.
Textual version (optional)	*When reversing in heavy weather conditions, the car's ECUs might be overloaded with computations and therefore not be able to handle all calculations related to the video feed processing.*

Fig. 6.16 Scenario of overloading of the main ECU

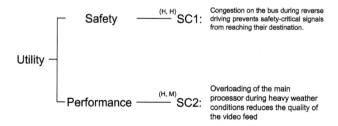

Fig. 6.17 Quality attribute utility tree

The reason for including both scenarios is the fact that they illustrate different possibilities of reasoning about deployment of functionality on nodes.

The quality attribute utility tree in our case consists of these two scenarios linked to two attributes—performance and safety. Both of these scenarios are ranked as high (H) in the utility tree, as shown in Fig. 6.17.

Now that we have the utility tree let us analyze the two architecture scenarios, and describe the trade-offs and sensitivity points.

6.5.5 Analysis of the Architecture and the Architectural Decision

Now we can analyze the architecture and its two deployments. In this analysis we can use a number of risks, for example the risk that the signal does not reach its destination. We can describe the risk using the template described in this chapter. The description is presented in Fig. 6.18.

Since the risk presented in Fig. 6.18 affects the safety of the passengers, it should be reduced. Reduction of this risk means that communication over the bus should not affect the safety-critical signals. Therefore the architectural decision is that priority should be given the deployment alternative—i.e. placing the processing of the video feed on the BBC ECU rather than on the main ECU.

The alternative means that the BBC ECU should have sufficient processing power to process the video in real time, which may increase the cost of the electrical components in the car. However, safety can allow the company to pursue its main business model (as described by the business drivers) and therefore balance the increased cost with increased sales of cars.

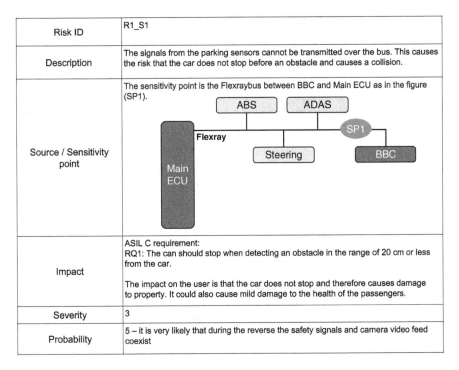

Fig. 6.18 Risk description

6.5.6   Summary of the Example

In this example we presented a simple assessment of a part of the software architecture for a car. The intention of this example is to provide an insight on how to think and reason when conducting such an assessment. In practice, the main purpose of an assessment like this one is all the discussions and presentations conducted by the assessment and the architecture teams. The questions, scenarios, prioritizations, and simply, brainstorming of ideas are the main point and benefit of the architecture. We summarize them in table presented in Fig. 6.19.

The ATAM procedure is defined for software architectures, but in the automotive domain the deployments of the software components and physical hardware architectures are tightly connected to the software—they both influence the software architecture and are influenced by the architecture (as this example assessment shows). Therefore, our advise is to always broaden the assessment team to include both software specialists and the hardware specialists—to cover the system properties of software architectures.

Scenario 5	Capture video during the reverse driving (backing up) of the car from the rear-camera and show it on the main display.		
Attributes	Safety.		
Environment	Car in reverse driving.		
Stimulus	Camera feed to be shown on the display.		
Response	Process video data and show it on the display.		
Architectural decisions	Sensitivity	Trade-off	Risk
Placing the processing of the video feed on the Main ECU	S1	T1	R1
Placing the processing of the video feed on BBC		T2	R2
Reasoning	The functioning of the main ECU is vital to the system (see sensitivity point S1) Safety versus lowered cost (see trade-off point T1) Safety requirement might be at risk due to heavy processing on Main ECU (see risk R1)		
Architecture diagram	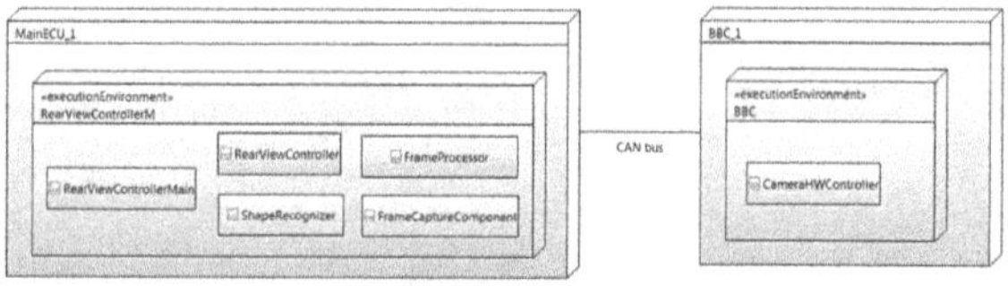		

Fig. 6.19 Tabular summary of the example ATAM evaluation

6.6 Further Reading

An interesting overview of scenario-based software architecture evaluation methods has been presented by Ionita et al. [IHO02]. Readers interested in a comparison between the methods are directed to this interesting article.

This article can be complemented by the work of Dobrica and Niemela [DN02], which focused on a more general overview and comparison of architecture evaluation methods.

A comprehensive work on the notion of graceful degradation has been presented by Shelton [She03, SK03] who discusses the notion of graceful degradation in the context of an example safety-critical system of an elevator, its modelling and measurement.

Readers interested in a wider view of the applicability of ATAM in other domains can look into the work of Bass et al. [BM$^+$01], who analyzed the architecture evaluation scenarios of a number of safety-critical systems.

The original works of Bass and Kazman have been expanded to other domains and other quality attributes than the original few (modifiability, reliability, availability). An example of such extensions is presented by Govseva et al. [GPT01] and Folmer and Bosch [FB04].

In the automotive domain we often consider different car models as product lines with the equipment levels as product line members. For this kind of view on automotive software architectures one could find the extension of ATAM to capture product lines to be interesting—[OM05].

Readers interested in further examples of architecture evaluations can be found in the article by Bergey et al. [BFJK99], who describe the experiences of using ATAM in the context of software acquisitions. The readers can also consider the work of Barbacci et al. [BCL$^+$03].

6.7 Summary

Architecting is a discipline of high-level design which is often described in the form of diagrams. However, equally important to the design is the set of decisions taken when creating the architecture. These decisions delineate a set of principles which designers have to follow in order to make sure that the software system fulfills its purpose.

Arriving at the right decisions is a process of combining the expertise of architects and the considerations of architects and designers. In this chapter we presented a method to elicit architectural decisions based on discussions between an external evaluation team and the architecture team—ATAM (Architecture Tradeoff Analysis Method). Through the assessments we can learn about the principles behind the architectural design and design decisions. We can learn about the alternative choices and why they are rejected.

In this chapter we focus on the "human" aspects of software architecture evaluation, which is by definition bound to be subjective to a certain degree. In the next chapter, however, we focus on the monitoring of the architecture quality given the set of information needs. This monitoring is done by conducting measurements and quantifying quality attributes discussed in this chapter.

References

A+08.   Motor Industry Software Reliability Association et al. *MISRA-C: 2004: guidelines for the use of the C language in critical systems.* MIRA, 2008.

BCL+03.   Mario Barbacci, Paul C Clements, Anthony Lattanze, Linda Northrop, and William Wood. Using the architecture tradeoff analysis method (atam) to evaluate the software architecture for a product line of avionics systems: A case study. 2003.

BFJK99.   John K Bergey, Matthew J Fisher, Lawrence G Jones, and Rick Kazman. Software architecture evaluation with atam in the dod system acquisition context. Technical report, DTIC Document, 1999.

BLBvV04.   PerOlof Bengtsson, Nico Lassing, Jan Bosch, and Hans van Vliet. Architecture-level modifiability analysis (alma). *Journal of Systems and Software*, 69(1):129–147, 2004.

BM+01.   Len Bass, Gabriel Moreno, et al. Applicability of general scenarios to the architecture tradeoff analysis method. Technical report, DTIC Document, 2001.

DN02.   Liliana Dobrica and Eila Niemela. A survey on software architecture analysis methods. *IEEE Transactions on software Engineering*, 28(7):638–653, 2002.

FB04.   Eelke Folmer and Jan Bosch. Architecting for usability: a survey. *Journal of systems and software*, 70(1):61–78, 2004.

GPT01.   Katerina Goševa-Popstojanova and Kishor S Trivedi. Architecture-based approach to reliability assessment of software systems.*Performance Evaluation*, 45(2):179–204, 2001.

IHO02.   Mugurel T Ionita, Dieter K Hammer, and Henk Obbink. Scenario-based software architecture evaluation methods: An overview. *Icse/Sara*, 2002.

ISO16a.   ISO/IEC. ISO/IEC 25000 - Systems and software engineering - Systems and software Quality Requirements and Evaluation (SQuaRE). Technical report, 2016.

ISO16b.   ISO/IEC. ISO/IEC 25023 - Systems and software engineering - Systems and software Quality Requirements and Evaluation (SQuaRE) - Measurement of system and software product quality. Technical report, 2016.

KKB+98.   Rick Kazman, Mark Klein, Mario Barbacci, Tom Longstaff, Howard Lipson, and Jeromy Carriere. The architecture tradeoff analysis method. In *Engineering of Complex Computer Systems, 1998. ICECCS'98. Proceedings. Fourth IEEE International Conference on*, pages 68–78. IEEE, 1998.

KKC00.   Rick Kazman, Mark Klein, and Paul Clements. Atam: Method for architecture evaluation. Technical report, DTIC Document, 2000.

LD97.   Oliver Laitenberger and Jean-Marc DeBaud. Perspective-based reading of code documents at robert bosch gmbh. *Information and Software Technology*, 39(11):781–791, 1997.

LSR07.   Frank Linden, Klaus Schmid, and Eelco Rommes. The product line engineering approach. *Software Product Lines in Action*, pages 3–20, 2007.

OC01.   International Standard Organization and International Electrotechnical Commission. Iso iec 9126, software engineering, product quality part: 1 quality model. Technical report, International Standard Organization / International Electrotechnical Commission, 2001.

OM05.   Femi G Olumofin and Vojislav B Misic. Extending the atam architecture evaluation to product line architectures. In *5th Working IEEE/IFIP Conference on Software Architecture (WICSA'05)*, pages 45–56. IEEE, 2005.

RSB+13.   Rakesh Rana, Miroslaw Staron, Claire Berger, Jorgen Hansson, Martin Nilsson, and Fredrik Torner. Evaluating long-term predictive power of standard reliability growth models on automotive systems. In *Software Reliability Engineering (ISSRE), 2013 IEEE 24th International Symposium on*, pages 228–237. IEEE, 2013.

RSB+16.   Rakesh Rana, Miroslaw Staron, Christian Berger, Jörgen Hansson, Martin Nilsson, and Wilhelm Meding. Analyzing defect inflow distribution and applying bayesian inference method for software defect prediction in large software projects. *Journal of Systems and Software*, 117:229–244, 2016.

RSM+13.   Rakesh Rana, Miroslaw Staron, Niklas Mellegård, Christian Berger, Jörgen Hansson, Martin Nilsson, and Fredrik Törner. Evaluation of standard reliability growth models in the context of automotive software systems. In *Product-Focused Software Process Improvement*, pages 324–329. Springer, 2013.

She03.   Charles Preston Shelton. *Scalable graceful degradation for distributed embedded systems*. PhD thesis, Carnegie Mellon University, 2003.

SK03.   Charles Shelton and Philip Koopman. Using architectural properties to model and measure graceful degradation. In *Architecting dependable systems*, pages 267–289. Springer, 2003.

SM16.   Miroslaw Staron and Wilhelm Meding. Mesram–a method for assessing robustness of measurement programs in large software development organizations and its industrial evaluation. *Journal of Systems and Software*, 113:76–100, 2016.

TRW03.   Thomas Thelin, Per Runeson, and Claes Wohlin. An experimental comparison of usage-based and checklist-based reading. *IEEE Transactions on Software Engineering*, 29(8):687–704, 2003.

Chapter 7
Metrics for Software Design and Architectures

Miroslaw Staron
University of Gothenburg, Gothenburg, Sweden

Wilhelm Meding
Ericsson AB, Gothenburg, Sweden

Abstract Understanding the architecture in a qualitative manner can be time-consuming and effort-intensive. Therefore the qualitative methods such as assessments presented in Chap. 6 are often done periodically at given milestones. However, architects need to monitor the quality of the architecture constantly and ensure that the characteristics of the architecture are within the limits of the product boundaries. In this chapter we present a set of measures used for measuring architectures and detailed designs. We explore the existing measures and present the ones which are common in industrial applications. Towards the end of the chapter we show the limits of selected measures by using an openly available industrial data set from an automotive OEM.

7.1 Introduction

In the previous chapter we explored one way of understanding the architecture—qualitative assessment based on scenarios. This method has multiple advantages as it allows architects to dive deeply into the details of a selected set of prioritized aspects of the architecture. The major disadvantage is the fact that qualitative evaluation is effort-intensive and can be done as soon as the architecture is somehow mature.

Architecting, however, is not done when the architecture is finished but is done intensively before the architecture is finished. Moreover, it is done constantly, so periodical assessments need to be complemented with methods for continuous quality assessment. In order to achieve this continuity we need to use automated methods which are usually based on measuring properties of architectures and properties of detailed designs.

© Springer International Publishing AG 2017
M. Staron, *Automotive Software Architectures*,
DOI 10.1007/978-3-319-58610-6_7

Software architecting as an area has gained increasing visibility in the last two decades as the software industry has recognized the role of software architectures in maintaining high quality and ensuring longevity and sustainability of software products [Sta15, LKM⁺13]. Even though this recognition is not new, there is still no consensus on how to measure various aspects of software architectures beyond the basic structural properties of the software architecture as a design artifact. In the literature we can encounter studies applying base measures for object-oriented designs to software architectures [LTC03] and studies designing low-level software architecture measures such as number of interfaces [SFGL07].

In order to understand the kinds of measures which are used in software architectures we have found a generic measurement portfolio of 54 measures in the literature. The portfolio can be applied to software architectures and designs, but interpreted differently based on where it is applied. The portfolio was developed by the literature review using snowballing and following the principles of systematic mapping of Petersen et al. [PFMM08]. The measures in the portfolio were then organized according to the ISO/IEC 15939 standard's measurement information model [OC07] into base measures, derived measures and indicators.

This chapter is structured as follows. Next, Sect. 7.2 presents our theoretical foundation for designing the portfolio—the ISO/IEC 15939 measurement information model. In Sect. 7.3 we present an overview of the standardized measures presented in the new quality standard "Software Product Quality Requirements and Evaluation". In Sect. 7.4 we present more measures found in literature and we organize them in the portfolio in Sect. 7.5 by identifying indicators. In Sect. 7.6 we present the limits of the selected measures based on an open data set from an automotive OEM. We conclude the chapter with further reading in Sect. 7.7.

7.2 Measurement Standard in Software Engineering: ISO/IEC 15939

The ISO/IEC 15939:2007 [OC07] standard is a normative specification for processes used to define, collect, and analyze quantitative data in software projects or organizations. The central role in the standard is played by the information product, which is a set of one or more indicators with their associated interpretations that address the information need. The information need is an insight necessary for a stakeholder to manage objectives, goals, risks, and problems observed in measured objects. These measured objects can be entities like projects, organizations, software products, etc. characterized by a set of attributes. We use the following definitions from ISO/IEC 15939:2007:

- Base measure, defined in terms of an attribute and the method for quantifying it. This definition is based on the definition of base quantity from [oWM93].
- Derived measure, defined as a function of two or more values of base measures. This definition is based on the definition of derived quantity from [oWM93].
- Indicator, provides an estimate or evaluation of specified attributes derived from a model with respect to defined information needs.
- Decision criteria—thresholds, targets, or patterns used to determine the need for action or further investigation, or to describe the level of confidence in a given result.
- Information product—one or more indicators and their associated interpretations that address an information need.
- Measurement method—a logical sequence or operations, described generically, used in quantifying an attribute with respect to a specified scale.
- Measurement function—an algorithm or calculation to combine two or more base measures.
- Attribute—a property or characteristic of an entity that can be distinguished quantitatively or qualitatively by human or automated means.
- Entity—an object that is to be characterized by measuring its attributes.
- Measurement process—a process for establishing, planning, performing and evaluating measurement within an overall project, enterprise or organizational measurement structure.
- Measurement instrument a procedure to assign a value to a base measure.

The view on measures presented in ISO/IEC 15939 is consistent with other engineering disciplines; the standard states at many places that it is based on such standards as ISO/IEC 15288:2007 (Software and Systems engineering—Measurement Processes), ISO/IEC 14598-1:1999 (Information technology—Software product evaluation), ISO/IEC 9126-x, the ISO/IEC 25000 series of standards, and the International vocabulary of basic and general terms in metrology (VIM) [oWM93]. Conceptually, the elements (different kinds of measures) which are used in the measurement process can be presented as in Fig. 7.1.

The model provides a very good abstraction and classification of measures—from very basic ones to more complicated ones. The base measures are often close to the entities they measure, such as architectural designs, and as such reflect the entities relatively well, although using a different domain of mathematical symbols and numbers. The indicators, on the other hand, serve the different

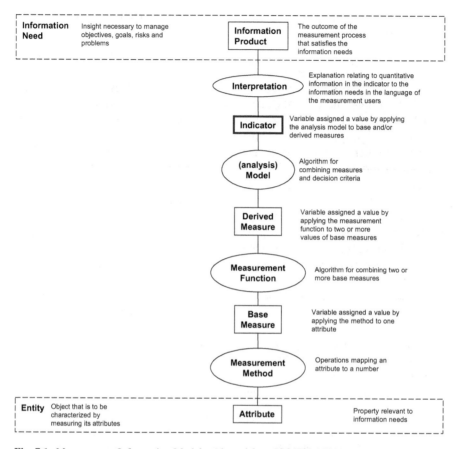

Fig. 7.1 Measurement Information Model—adopted from ISO/IEC 15939

purpose of fulfilling the information need of their stakeholder and as such are closer to the concepts which the stakeholders want to get information about, e.g. the architecture's quality, stability or complexity.

As the indicators provide insight into what the stakeholders would like to measure, see and observe, it is often easy to provide an analysis model (or coloring) of the values of the indicators. It can be illustrated as in Fig. 7.2.

We use this model to describe the measures used for quantifying properties of software architectures. Conceptually we can also consider the fact the higher in the model the measure is, the more advanced the information need it fulfills. In Fig. 7.3 we can see a number of measures divided into three levels—the more basic ones at the bottom and the more complex ones at the top.

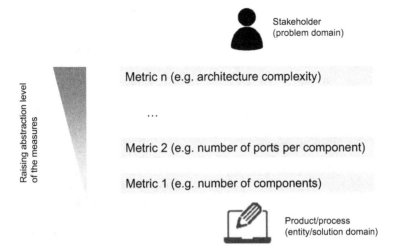

Fig. 7.2 Conceptual levels of architecture measures

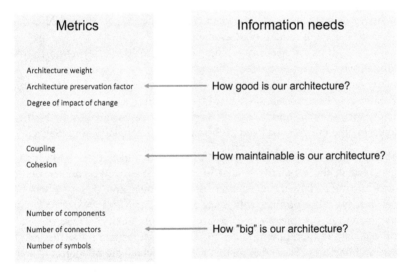

Fig. 7.3 Higher-level measures correspond to more advanced information needs—an example

The more advanced information needs are related to the work of the architects whereas the more basic ones are more related to the architecture as an artifact in software development. So, now that we have the model, let's look into one of the standards where the software measures are defined—ISO/IEC 25000.

7.3   Measures Available in ISO/IEC 25000

The ISO/IEC 25000 Software Quality Requirements and Evaluation (SQuaRE) standard provides a set of reference measures for software designs and architectures. At the time of writing of this book the standard is not fully adopted but the main parts are already approved and the work is fully ongoing regarding the measures, their definitions and usage. The standard presents the following set of measures related to product, design and architecture in one of its chapters—ISO/IEC 25023—Software and Software Product Quality Measures [ISO16]:

- Quality measures for functional suitability—example measure: functional implementation coverage addressing the information need of functional completeness
- Quality measures for performance efficiency—example measure: response time addressing the information need of time behavior performance
- Quality measures for compatibility—example measure: connectivity with external systems addressing the information need of interoperability
- Quality measures for usability—example measure: completeness of user documentation addressing the information need of learnability of the product
- Quality measures for reliability—example measure: test coverage addressing the information need of reliability assessment
- Quality measures for security—example measure: data corruption prevention addressing the information need of integrity
- Quality measures for maintainability—example measure: modification complexity addressing the information need of modifiability
- Quality measures for portability—example measure: installation time efficiency addressing the information need of installability of the software product

The list of the areas and the example measures illustrate how the measures are discussed in the standards related to product quality. We can see that these measures are related to the execution of the product and do not focus on the internal quality of the product with such example measures as size (e.g. number of components) or complexity (e.g. control flow complexity). Therefore we need to turn to scientific literature to understand the measures and indicators related to software architectures. There we can find measures which are of interest to software architects.

7.4   Measures

Let's start with the base measures which quantify the architecture shows in Table 7.1—we can quickly notice that these measures correspond to the entities they measure. The measurement method (the algorithms to calculate the base measure) are very similar and are based on counting entities of a specific type. The list in Table 7.1 shows a set of example base measures.

Table 7.1 Base measures for software architectures

Measure	Description
Number of components [SJZ14]	The basic measure quantifying the size of the architecture in terms of its basic building block—components
Number of connectors [SJZ14]	The basic measure quantifying the internal connectivity of the architecture in terms of its basic connectors
Number of processing units [LK00]	The basic measure quantifying the size of the physical architecture in terms of processing units
Number of data repositories [LK00]	The complementary measure quantifying the size in terms of data repositories
Number of persistent components [LK00]	Quantifies the size in terms of the need for persistency
Number of links [LK00]	Quantifies the complexity of the architecture, similarly to the McCabe cyclomatic complexity measure. It is sometimes broken down by type of link (e.g, asynchronous–synchronous, data-control)
Number of types of communication mechanisms [LK00]	Quantifies the complexity of the architecture in terms of the need to implement multiple communication mechanisms
Number of external interfaces [KPS+98]	Quantifies the coupling between architectural components and external systems
Number of internal interfaces [KPS+98]	Quantifies the coupling among the architectural components
Number of services [KPS+98]	Quantifies the cohesion of the architecture in terms of how many services it provides/fulfills
Number of concurrent components [KPS+98]	The measure counts the components which have concurrent calculations as part of their behavior
Number of changes in the architecture [DNSH13]	The measure quantifies the number of changes (e.g. changed classes, changed attributes) in the architecture
Fanout from the simplest structure [DSN11]	The measure quantifies the degree of the lowest complexity of the coupling of the architecture

Collecting the measures presented in the table provides the architects with the understanding of the properties of the architecture, but the architects still need to provide context to these numbers in order to reason about the architectures. For example, the number of components by itself does not provide much insight; however, if put together with a timeline and plotted as a trend, allow to extrapolate the information and therefore allow the architects to assess if the architecture is overly large and should be refactored.

Table 7.2 Base measures for software design

Measure	Description
Weighted methods per class [CK94]	The number of methods weighed by their complexity
Depth of inheritance tree [CK94]	The longest path from the current class to its first predecessor in the inheritance hierarchy
Cyclomatic complexity [McC76]	Quantifies the control path complexity in terms of the number of independent execution paths of a program. Used often as part of safety assessment in ISO/IEC 26262
Dependencies between blocks/modules/classes [SMHH13]	Quantifies the dependencies between classes or components in the system
Abstractness of a Simulink block [Ols11]	Quantifies the ratio of contained abstract blocks to the total number of contained blocks

In addition to the measures for the architecture we can also find many measures which are related to software design in general—e.g. object-oriented measures or complexity measures [ASM+14, SKW04]. Examples of these are presented in Table 7.2.

Once again these examples show that the measures are related to the design the quantification of its properties. Such measures as the *abstractness of a Simulink block*, however, are composed of multiple other measures and therefore are classified as derived measures and as such are closer to the information need of architects. In the literature we can find a large number of measures for designs and their combinations and therefore when choosing measures it is crucial to start from the information needs of the architects [SMKN10] since these information needs can effectively filter out measures which are possible to collect, but not relevant for the company (and as such could be considered as waste).

In the next section we identify which measures from the above two groups are to be included in the portfolio and what areas they belong to.

7.5   Metrics Portfolio for the Architects

The measures presented so far can be collected, but, as the measurement standards prescribe, they need to be useful for the stakeholders in their decision processes [Sta12, OC07]. Therefore we organize these measures into three areas corresponding to the information needs of software architects. As architecting is a process which involves software architecture artifacts, we recognize the need of grouping these indicators into areas related to both the product and the process.

7.5.1 Areas

In our portfolio we group the indicators into three areas related to basic properties of the design, its stability and its quality:

Area: architecture measures—this area groups product-related indicators that address the information need about *how to monitor the basic properties of the architecture, like its component coupling.*

Area: design stability—this area groups process-related indicators that address the information need about *how to ascertain controlled evolution of the architectural design.*

Area: technical debt/risk—this area groups product-related indicators that address the information need about *how to ascertain the correct implementation of the architecture.*

In the following subsections we present the measures and the suggested way to present them. One of the criteria for each of these areas in our study was that the upper limit on the number of indicators be four. The limitations are based on empirical studies of cognitive aspects of measurement, such as the ability to take in information by the stakeholders [SMH+13].

7.5.2 Area: Architecture Measures

In our portfolio we could identify 14 measures as applicable to measure the basic properties of the architecture. However, when discussing these measures with the architects, the majority of the measures seemed to quantify basic properties of the designs. The indicators found in the study in this area are:

Software architecture changes: To monitor and control changes over time the architects should be able to monitor the trends in changes of software architecture at the highest level [DNSH13]. Based on our literature studies and discussions with practitioners we identified the following measure to be a good indicator of the changes—*number of changes in the architecture per time unit (e.g. week)* [DSTH14a, DSTH14b, DSN11].

Complexity: To manage module complexity, the architects need to understand the degree of coupling between components, as the coupling is perceived as cost-consuming and error-prone in the long-term evolution of the architecture. The identified indicator is *Average squared deviation of actual fanout from the simplest structure.*

External interfaces: To control the degree of coupling on the interface level (i.e. a subset of all types of couplings), the architects need to observe the number of internal interfaces—*number of interfaces*.

Internal interfaces: To control of external dependencies of the product, the architects need to monitor the coupling of the product to external software products—*number of interfaces*.

The suggested presentation of these measures is presented in Fig. 7.4.

7.5.3 Area: Design Stability

The next area which is of importance for the architects is related to the need for monitoring the large code base for stability. Generally, in this area we used visualizations from our previous research into code stability [SHF+13]. We identified the following three indicators to be efficient in monitoring and visualizing the stability:

Code stability: To monitor the code maturity over the time the architects need to see how much code has been changed over time as it allows them to identify code areas where more testing is needed due to recent changes. The measure used for this purpose is *number of changes per module per time unit*.

Defects per modules: To monitor the aging of the code the architects need to monitor defect-proneness per component per time, using a similar measure as that for code stability—*number of defects per module per time unit (e.g. week)*.

Interface stability: To control the stability of the architecture over its interfaces the architects measure the stability of the interfaces—*number of changes to the interfaces per time unit*.

We have found that it is important to be able to visualize the entire code/product base in one view and therefore the dashboard which depicts the stability is based on the notion of heatmaps [SHF+13]. In Fig. 7.5 we present such a visualization with three heatmaps corresponding to these three stability indicators. Each of the figures is a heatmap which depicts different aspects, but each of them is organized in the same way—columns designate weeks, rows designate the single code modules or interfaces and the intensity of the color of each cell designates the number of changes to the module or interface during the particular week.

7.5.4 Area: Technical Debt/Risk

The last area in our portfolio is related to the quality of the architecture over a longer period of time. In this area we identified the following two indicators:

Coupling: To have manageable design complexity the architects need to have a way to get a quick overview over the coupling between the components in the

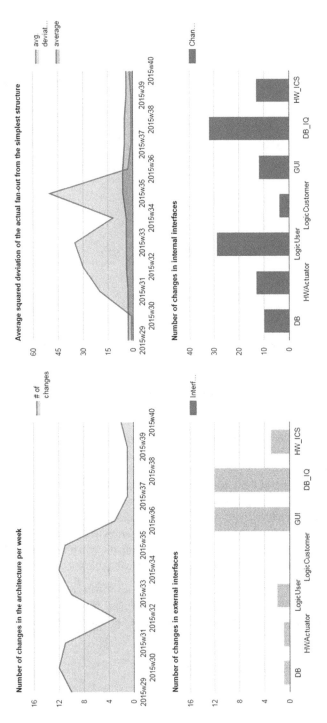

Fig. 7.4 Visualization of the measures in the architecture property area

Code stability heatmap

Defects per module heatmap

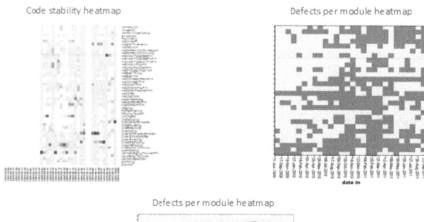

Defects per module heatmap

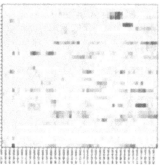

Fig. 7.5 Visualization of the measures in the architecture stability area

architecture—measured by *number of explicit architectural dependencies*, where the explicit dependencies are links between the components which are introduced by the architects.

Implicit architectural dependencies: To monitor where the code deviates from the architecture the architects need to observe whether there are any additional dependencies introduced during the detailed design of the software—this is measured by *number of implicit architectural dependencies*, where the implicit dependencies are such links between the components which are part of the code, but not introduced in the architecture documentation diagrams [SMHH13].

The visualization of the architectural dependencies shows the degree of coupling and is based on circular diagrams, as presented in Figs. 7.6 and 7.7, where each area on the border of the circle represents a component and a line shows a dependency between two components.

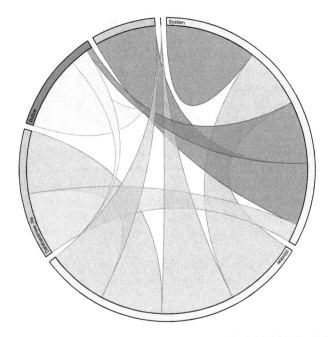

Fig. 7.6 Visualization of the measures in the architecture technical debt/risk: implicit

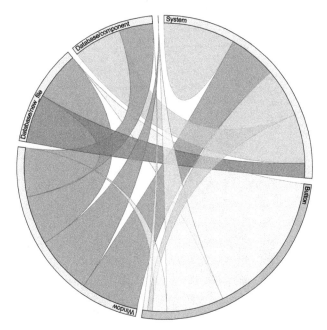

Fig. 7.7 Visualization of the measures in the architecture technical debt/risk: explicit

7.6 Industrial Measurement Data for Software Designs

The metrics portfolio for software architects should be complemented with a set of metrics for software designs, which we presented in Table 7.2. One of these measures is software complexity, measured as a number of independent paths in the program (McCabe complexity). In order to illustrate how complex automotive systems are, let us look into one of the industrial data sets publicly available [ASD⁺15].

In general, software complexity can be measured in multiple ways, but there is a small number of measures which have been found to be correlated with each other—e.g. McCabe cyclomatic complexity, lines-of-code. The inherent correlations (cf. [ASH⁺14]) allow us to simplify the problem to only one of them (for the sake of the discussion)—we choose the McCabe complexity due to its spread in practice. In short, the metric measures the number of independent execution paths in the source code.

In the automotive sector, in data from the open domain we find that the complexity of software modules is highly over the theoretical limit of 30 (execution paths), as presented in Fig. 7.8.

What the data shows is that there are components where the number of execution paths is over 160, which means that only to test each of the execution paths once there is a need for 160+ test cases. However, in order to achieve full coverage one needs more than 500 test cases for the entire component. If we need to test each path with a positive and a negative case (so called boundary case) we need to at least double the number of test cases. Exploring the other metrics provided in the

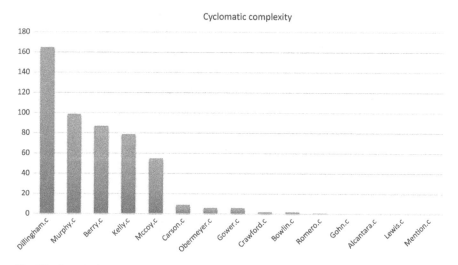

Fig. 7.8 Complexity of software modules (C programming language) as a McCabe cyclomatic complexity

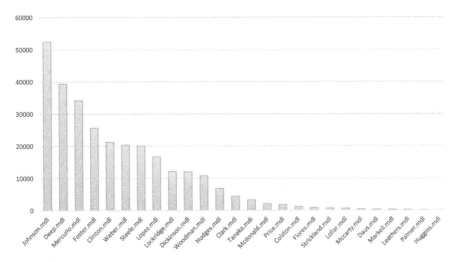

Fig. 7.9 Sizes of the models in the example data set

same data set shows that the trends are very similar—the numbers are highly over the theoretical complexity limits.

These numbers indicate that it is increasingly more difficult to provide full verification of the software functionality in order to ensure the safety of software systems. Therefore we need new approaches than just testing.

In Chap. 5 we explored the detailed designs in terms of simulink models. In the data set presented in the studied paper [ASD+15], the size of such models can be huge—as shown in Fig. 7.9.

As the figure shows, some of the models (Johnson.mdl) are huge models with over 50,000 blocks and models of over 10,000 blocks are not uncommon. One should note that this data comes only from one domain and one manufacturer; however, the scale of the size shows how much software is included in modern cars. It also shows the effort required to develop and to test such software.

7.7 Further Reading

Some of the most popular methods for evaluating software architectures in general are to use qualitative methods like ATAM [KKC00], where the architecture is analyzed based on scenarios or perspectives. These methods are used for final assessments of the quality of the architectures, but as they are manual they need effort and therefore cannot be conducted in a continuous manner. However, as many contemporary projects are conducted using Agile methodologies, Lean software development [Pop07] or the minimum viable product approach [Rie11], these methods are not feasible in practice. Therefore the architects are willing to trade off quality of evaluation for speed of the feedback on their architecture, which leads to more extensive use of measure-based evaluation of software architectures.

In our previous work we have studied metrics used for monitoring of architectural changes [DNSH13, DSN11]. The results showed that the use of a modified coupling metric can provide a very good estimation of the impact of the change in the architecture between two different releases of the architecture.

One of the tools and methods supporting the architects' work with measures is the MetricViewer [TLTC05], which augments software architecture diagrams expressed in UML with such measures as coupling, cohesion and depth of inheritance tree. This augmentation is important for reasoning about the designs, but it is not linked to the information needs of the stakeholders. Having such a link allows the stakeholders to monitor attainment of their goals, which otherwise require them to conduct the same analyses manually.

Similarly to Tameer et al., Vasconcelos et al. [VST07] propose a set of metrics for measuring architectures based on low-level properties of software architectures, such as number of possible operating systems or number of secure components. Our work complements their study by focusing on internal quality properties related to the design and not quality in use.

In the same vein, Dave [Dav01] patented the method for co-synthesis of software and hardware using measures such as scheduling and task allocation metrics, which complement the portfolio of architecture metrics presented in this chapter. The major difference in the approach of the patent and our research is our focus on three areas and their associated information needs rather than on a specific goal–integration.

Additionally, even though it is a decade old, the technical recommendation for the architecture evaluation still provides useful guidelines for choosing the right method [ABC+97]. In particular, the recommendation is to customize the evaluation to a specific quality or goal. In the case of the study presented in this chapter, this goal is the set of information needs represented by the stakeholder.

The specific view, information need or goal which is prescribed in the architecture evaluation is a specific case of the *domain context* of the metrological properties of measures [Abr10]. In software engineering in general and in software architectures in particular there is no consensus about the universal values of measures (e.g. how strongly coupled two entities should be), and therefore the stakeholders approximate this using their experience and mandate in product development organizations [RSB+13, RSB+14, RSM+13].

Readers interested in other examples of information needs for software metrics are referred to a survey study conducted at Microsoft where the authors interview over 100 engineers, managers, and testers to map their current and future information needs [BZ12].

Using business intelligence and corporate performance measurement can be of interest to readers interested in decision making at the strategic level, e.g., [Pal07, RW01, KN98].

Readers interested in mechanisms of effective visualization and manipulation of measurement data can explore the field of visual analytics, e.g., [VT07, Tel14, BOH11].

Close to the field of visual analytics is the field of project telemetry, which focuses on online visualization of selected software metrics; interested readers should explore:

- tools like Hackystat that are examples in this field [Joh01, JKA$^+$03]
- the SonarQube tool suite for monitoring internal quality of software products during development [HMK10] and
- dashboards for visualizing product development where the authors describe experiences from introducing dashboards for a single team [FSHL13].

Readers interested in the concepts of measurement systems should explore the following publications:

- ISO/IEC 15939 (and its IEEE correspondent), defining the concepts related to measurement systems [OC07].
- Practical Software Measurement [McG02].
- The classical book on software metrics by Fenton and Pfleeger [FB14].
- The process of designing measurement systems in industry [SMN08].
- The graphical way of designing measurement systems with the focus on the information need of the stakeholders [SM09].

One of the trends observed in the software industry is the growing focus on customers even in measurement of internal quality attributes. Readers interested in how to work with customer data can find the following works of value:

- Post-deployment data [OB13],
- developing customer profiles [AT01], and
- mining and visualizing customer data [Kei02].

In this context of customer data collections, it is also important to understand the defects in automotive software. In our previous work we have developed a method for classifying defects based on their criticality, targeted towards automotive software [MST12] which is related to studies on the understanding of inconsistencies in designs [KS03].

7.8 Summary

In this chapter we focused on the challenge of constantly monitoring the architecture quality and the properties of software designs. We have focused on two aspects—what measures exist in the literature that can be used for this purpose and which measures should be used as indicators.

In the chapter we used the approach postulated by modern measurement standards in software engineering—ISO/IEC 15939 and ISO/IEC 25000. The first of this standard provided us with a way of structuring the measures and the second one provided us with a list of measures. Based on our work with industrial partners [SM16] we identified three areas of interest. In these areas we managed to identify a set of measures and indicators which address the needs of the stakeholders.

Finally we have also presented reference visualizations of these indicators.

References

ABC+97.   Gregory Abowd, Len Bass, Paul Clements, Rick Kazman, and Linda Northrop. Recommended best industrial practice for software architecture evaluation. Technical report, DTIC Document, 1997.

Abr10.   Alain Abran. *Software metrics and software metrology*. John Wiley & Sons, 2010.

ASD+15.   Harry Altinger, Sebastian Siegl, Dajsuren, Yanja, and Franz Wotawa. A novel industry grade dataset for fault prediction based on model-driven developed automotive embedded software. In *12th Working Conference on Mining Software Repositories (MSR)*. MSR 2015, 2015.

ASH+14.   Vard Antinyan, Miroslaw Staron, Jörgen Hansson, Wilhelm Meding, Per Osterström, and Anders Henriksson. Monitoring evolution of code complexity and magnitude of changes. *Acta Cybernetica*, 21(3):367–382, 2014.

ASM+14.   Vard Antinyan, Miroslaw Staron, Wilhelm Meding, Per Österström, Erik Wikstrom, Johan Wranker, Anders Henriksson, and Jörgen Hansson. Identifying risky areas of software code in agile/lean software development: An industrial experience report. In *Software Maintenance, Reengineering and Reverse Engineering (CSMR-WCRE), 2014 Software Evolution Week-IEEE Conference*, pages 154–163. IEEE, 2014.

AT01.   Gediminas Adomavicius and Alexander Tuzhilin. Using data mining methods to build customer profiles. *Computer*, 34(2):74–82, 2001.

BOH11.   Michael Bostock, Vadim Ogievetsky, and Jeffrey Heer. D^3 data-driven documents. *IEEE transactions on visualization and computer graphics*, 17(12):2301–2309, 2011.

BZ12.   Raymond PL Buse and Thomas Zimmermann. Information needs for software development analytics. In *Proceedings of the 34th international conference on software engineering*, pages 987–996. IEEE Press, 2012.

CK94.   Shyam R Chidamber and Chris F Kemerer. A metrics suite for object oriented design. *Software Engineering, IEEE Transactions on*, 20(6):476–493, 1994.

Dav01.   Bharat P Dave. Hardware-software co-synthesis of embedded system architectures using quality of architecture metrics, January 23 2001. US Patent 6,178,542.

DNSH13.   Darko Durisic, Martin Nilsson, Miroslaw Staron, and Jörgen Hansson. Measuring the impact of changes to the complexity and coupling properties of automotive software systems. *Journal of Systems and Software*, 86(5):1275–1293, 2013.

DSN11.   Darko Durisic, Miroslaw Staron, and Martin Nilsson. Measuring the size of changes in automotive software systems and their impact on product quality. In *Proceedings of the 12th International Conference on Product Focused Software Development and Process Improvement*, pages 10–13. ACM, 2011.

DSTH14a.   Darko Durisic, Miroslaw Staron, Milan Tichy, and Jorgen Hansson. Evolution of long-term industrial meta-models–an automotive case study of autosar. In *Software Engineering and Advanced Applications (SEAA), 2014 40th EUROMICRO Conference on*, pages 141–148. IEEE, 2014.

DSTH14b.   Darko Durisic, Miroslaw Staron, Milan Tichy, and Jorgen Hansson. Quantifying long-term evolution of industrial meta-models-a case study. In *Software Measurement and the International Conference on Software Process and Product Measurement*

(IWSM-MENSURA), 2014 Joint Conference of the International Workshop on, pages 104–113. IEEE, 2014.

FB14.   Norman Fenton and James Bieman. *Software metrics: a rigorous and practical approach*. CRC Press, 2014.

FSHL13.   Robert Feldt, Miroslaw Staron, Erika Hult, and Thomas Liljegren. Supporting software decision meetings: Heatmaps for visualising test and code measurements. In *Software Engineering and Advanced Applications (SEAA), 2013 39th EUROMICRO Conference on*, pages 62–69. IEEE, 2013.

HMK10.   Hiroaki Hashiura, Saeko Matsuura, and Seiichi Komiya. A tool for diagnosing the quality of java program and a method for its effective utilization in education. In *Proceedings of the 9th WSEAS international conference on Applications of computer engineering*, pages 276–282. World Scientific and Engineering Academy and Society (WSEAS), 2010.

ISO16.   ISO/IEC. ISO/IEC 25023 - Systems and software engineering - Systems and software Quality Requirements and Evaluation (SQuaRE) - Measurement of system and software product quality. Technical report, 2016.

JKA+03.   Philip M Johnson, Hongbing Kou, Joy Agustin, Christopher Chan, Carleton Moore, Jitender Miglani, Shenyan Zhen, and William EJ Doane. Beyond the personal software process: Metrics collection and analysis for the differently disciplined. In *Proceedings of the 25th international Conference on Software Engineering*, pages 641–646. IEEE Computer Society, 2003.

Joh01.   Philip M Johnson. Project hackystat: Accelerating adoption of empirically guided software development through non-disruptive, developer-centric, in-process data collection and analysis. *Department of Information and Computer Sciences, University of Hawaii*, 22, 2001.

Kei02.   Daniel A Keim. Information visualization and visual data mining. *IEEE transactions on Visualization and Computer Graphics*, 8(1):1–8, 2002.

KKC00.   Rick Kazman, Mark Klein, and Paul Clements. Atam: Method for architecture evaluation. Technical report, DTIC Document, 2000.

KN98.   Robert S Kaplan and DP Norton. Harvard business review on measuring corporate performance. *Harvard Business School Press, EUA*, 1998.

KPS+98.   S Kalyanasundaram, K Ponnambalam, A Singh, BJ Stacey, and R Munikoti. Metrics for software architecture: a case study in the telecommunication domain. In *Electrical and Computer Engineering, 1998. IEEE Canadian Conference on*, volume 2, pages 715–718. IEEE, 1998.

KS03.   Ludwik Kuzniarz and Miroslaw Staron. Inconsistencies in student designs. In *the Proceedings of The 2nd Workshop on Consistency Problems in UML-based Software Development, San Francisco, CA*, pages 9–18, 2003.

LK00.   Chung-Horng Lung and Kalai Kalaichelvan. An approach to quantitative software architecture sensitivity analysis. *International Journal of Software Engineering and Knowledge Engineering*, 10(01):97–114, 2000.

LKM+13.   Patricia Lago, Rick Kazman, Niklaus Meyer, Maurizio Morisio, Hausi A Müller, and Frances Paulisch. Exploring initial challenges for green software engineering: summary of the first greens workshop, at icse 2012. *ACM SIGSOFT Software Engineering Notes*, 38(1):31–33, 2013.

LTC03.   Mikael Lindvall, Roseanne Tesoriero Tvedt, and Patricia Costa. An empirically-based process for software architecture evaluation. *Empirical Software Engineering*, 8(1):83–108, 2003.

McC76.   Thomas J McCabe. A complexity measure. *Software Engineering, IEEE Transactions on*, (4):308–320, 1976.

McG02.   John McGarry. *Practical software measurement: objective information for decision makers*. Addison-Wesley Professional, 2002.

MST12.   Niklas Mellegård, Miroslaw Staron, and Fredrik Törner. A light-weight software
            defect classification scheme for embedded automotive software and its initial evalu-
            ation. *Proceedings of the ISSRE 2012*, 2012.

OB13.   Helena Holmström Olsson and Jan Bosch. Towards data-driven product development:
            A multiple case study on post-deployment data usage in software-intensive embedded
            systems. In *Lean Enterprise Software and Systems*, pages 152–164. Springer, 2013.

OC07.   International Standard Organization and International Electrotechnical Commission.
            Software and systems engineering, software measurement process. Technical report,
            ISO/IEC, 2007.

Ols11.   Marta Olszewska. Simulink-specific design quality metrics. *Turku Centre for
            Computer Science*, 2011.

oWM93.   International Bureau of Weights and Measures. *International vocabulary of basic and
            general terms in metrology*. International Organization for Standardization, Genève,
            Switzerland, 2nd edition, 1993.

Pal07.   Bob Paladino. Five key principles of corporate performance management. *CMA
            MANAGEMENT*, 81(8):17, 2007.

PFMM08.   Kai Petersen, Robert Feldt, Shahid Mujtaba, and Michael Mattsson. Systematic
            mapping studies in software engineering. In *12th international conference on
            evaluation and assessment in software engineering*, volume 17, pages 1–10. sn, 2008.

Pop07.   Mary Poppendieck. Lean software development. In *Companion to the proceedings
            of the 29th International Conference on Software Engineering*, pages 165–166. IEEE
            Computer Society, 2007.

Rie11.   Eric Ries. *The lean startup: How today's entrepreneurs use continuous innovation to
            create radically successful businesses*. Random House LLC, 2011.

RSB+13.   Rakesh Rana, Miroslaw Staron, Christian Berger, Jörgen Hansson, Martin Nilsson,
            and Fredrik Törner. Increasing efficiency of iso 26262 verification and validation by
            combining fault injection and mutation testing with model based development. In
            ICSOFT, pages 251–257, 2013.

RSB+14.   Rakesh Rana, Miroslaw Staron, Christian Berger, Jörgen Hansson, Martin Nilsson,
            Fredrik Törner, Wilhelm Meding, and Christoffer Höglund. Selecting software
            reliability growth models and improving their predictive accuracy using historical
            projects data. *Journal of Systems and Software*, 98:59–78, 2014.

RSM+13.   Rakesh Rana, Miroslaw Staron, Niklas Mellegård, Christian Berger, Jörgen Hansson,
            Martin Nilsson, and Fredrik Törner. Evaluation of standard reliability growth models
            in the context of automotive software systems. In *Product-Focused Software Process
            Improvement*, pages 324–329. Springer, 2013.

RW01.   R Ricardo and D Wade. Corporate performance management: How to build a better
            organization through measurement driven strategies alignment, 2001.

SFGL07.   Cláudio Sant'Anna, Eduardo Figueiredo, Alessandro Garcia, and Carlos JP Lucena.
            On the modularity of software architectures: A concern-driven measurement frame-
            work. In *Software Architecture*, pages 207–224. Springer, 2007.

SHF+13.   Miroslaw Staron, Jorgen Hansson, Robert Feldt, Anders Henriksson, Wilhelm Meding,
            Sven Nilsson, and Christoffer Hoglund. Measuring and visualizing code stability –
            A case study at three companies. In *Software Measurement and the 2013 Eighth
            International Conference on Software Process and Product Measurement (IWSM-
            MENSURA), 2013 Joint Conference of the 23rd International Workshop on*, pages
            191–200. IEEE, 2013.

SJZ14.   Srdjan Stevanetic, Muhammad Atif Javed, and Uwe Zdun. Empirical evaluation of the
            understandability of architectural component diagrams. In *Proceedings of the WICSA
            2014 Companion Volume*, page 4. ACM, 2014.

SKW04.   Miroslaw Staron, Ludwik Kuzniarz, and Ludwik Wallin. Case study on a process
            of industrial MDA realization: Determinants of effectiveness. *Nordic Journal of
            Computing*, 11(3):254–278, 2004.

SM09.   Miroslaw Staron and Wilhelm Meding. Using models to develop measurement sys-
           tems: a method and its industrial use. In *Software Process and Product Measurement*,
           pages 212–226. Springer, 2009.
SM16.   Miroslaw Staron and Wilhelm Meding. A portfolio of internal quality measures for
           software architects. In *Software Quality Days*, pages 1–16. Springer, 2016.
SMH+13.   Miroslaw Staron, Wilhelm Meding, Jörgen Hansson, Christoffer Höglund, Kent
           Niesel, and Vilhelm Bergmann. Dashboards for continuous monitoring of quality
           for software product under development. *System Qualities and Software Architecture
           (SQSA)*, 2013.
SMHH13.   Miroslaw Staron, Wilhelm Meding, Christoffer Hoglund, and Jorgen Hansson. Identi-
           fying implicit architectural dependencies using measures of source code change waves.
           In *Software Engineering and Advanced Applications (SEAA), 2013 39th EUROMICRO
           Conference on*, pages 325–332. IEEE, 2013.
SMKN10.   M. Staron, W. Meding, G. Karlsson, and C. Nilsson. Developing measurement
           systems: an industrial case study. *Journal of Software Maintenance and Evolution:
           Research and Practice*, page 89–107, 2010.
SMN08.   Miroslaw Staron, Wilhelm Meding, and Christer Nilsson. A framework for devel-
           oping measurement systems and its industrial evaluation. *Information and Software
           Technology*, 51(4):721–737, 2008.
Sta12.   Miroslaw Staron. Critical role of measures in decision processes: Managerial
           and technical measures in the context of large software development organizations.
           Information and Software Technology, 54(8):887–899, 2012.
Sta15.   Miroslaw Staron. Software engineering in low-to middle-income countries. *Knowl-
           edge for a Sustainable World: A Southern African-Nordic contribution*, page 139,
           2015.
Tel14.   Alexandru C Telea. *Data visualization: principles and practice*. CRC Press, 2014.
TLTC05.   Maurice Termeer, Christian FJ Lange, Alexandru Telea, and Michel RV Chaudron.
           Visual exploration of combined architectural and metric information. In *Visualizing
           Software for Understanding and Analysis, 2005. VISSOFT 2005. 3rd IEEE Interna-
           tional Workshop on*, pages 1–6. IEEE, 2005.
VST07.   André Vasconcelos, Pedro Sousa, and José Tribolet. Information system architecture
           metrics: an enterprise engineering evaluation approach. *The Electronic Journal
           Information Systems Evaluation*, 10(1):91–122, 2007.
VT07.   Lucian Voinea and Alexandru Telea. Visual data mining and analysis of software
           repositories. *Computers & Graphics*, 31(3):410–428, 2007.

Chapter 8
Functional Safety of Automotive Software

Per Johannessen

Abstract In the previous chapters we explored generic methods for assessing quality of software architecture and software design. In this chapter we continue with a much-related topic, functional safety of software, in which functional safety assessment is one of the last activities during product development. We describe how the automotive industry works with functional safety. Much of this work is based on the ISO 26262 standard that was published in 2011. This version of the standard is applicable for passenger cars up to 3500 kg. There is also ongoing work on a future version, expected in 2018, applicable to most road vehicles, including buses, motorcycles, and trucks. The scope of the ISO 26262 standard is more than software development and for better understanding we give an overview of these other development phases in this chapter. However, we focus on software development according to ISO 26262. The different phases that are covered are software planning, software safety requirements, software architectural design, software unit design and implementation, software integration and testing, and verification of software.

8.1 Introduction

Functional safety in ISO 26262 is defined as *"absence of unreasonable risk due to hazards caused by malfunctioning behaviour of E/E systems"*. In a simplified way, we could say that there shall not be any harm to persons resulting from faults in electronics or software. At the same time, for an automotive product, this electronic and software is within a vehicle. Hence, when working with functional safety, it is important to consider the vehicle, the traffic situations including other vehicles and road users and the persons involved.

The safety lifecycle of ISO 26262 starts with planning of product development, continues with product development, production, operation and ends with scrapping the vehicle. In ISO 26262, the base for product development is Items. An Item in ISO 26262 is defined as "system or array of systems to implement a function at the vehicle level, to which ISO 26262 is applied". The key term here is "function at the vehicle level", which defines what components are involved. This also implies that a vehicle consists of many Items, to which ISO 26262 is applied.

The work on the ISO 26262 standard started in Germany in the early 2000 and was based on another standard, IEC 61508 ()Functional Safety of Electri-

© Springer International Publishing AG 2017

M. Staron, *Automotive Software Architectures*,

DOI 10.1007/978-3-319-58610-6_8

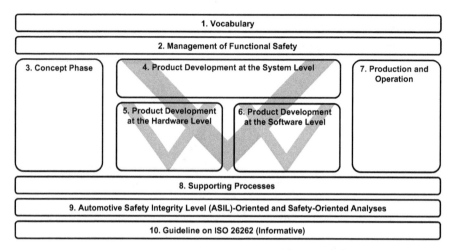

Fig. 8.1 The ten different parts in the ISO 26262 standard, adopted from [ISO11]

cal/Electronic/Programmable Electronic Safety-related Systems). As IEC 61508 [IEC10] originates from the process control industry, there was a need to adapt it to the automotive industry. The work within the ISO standardization organization started in 2005 and resulted in the first edition of ISO 26262 published in 2011.

Even if the automotive industry had been working with functional safety before, this was a significant step to standardize the work across the industry. As with standards in general, the key advantage is simplified cooperation between different organizations. Another benefit with ISO 26262 is that it can be seen as a cookbook on how to develop safe functions on the vehicle level that to some degree are implemented in electronics and software. By following this cookbook, the result is a harmonized safety level across the industry and this level is seen as acceptable.

When looking into ISO 26262, there are ten different parts as shown in Fig. 8.1. In this chapter we focus on part 6, for software development. At the same time, it is important to understand the context in which this software is developed and also the context in which this software is used. Hence, there is a very brief overview of these other parts in ISO 26262 as well.

As we can see in Fig. 8.1, parts 4–6 are based on the V-model of product development which we discussed in Chap. 3 and which is a defacto standard in the automotive industry. It should be noted that even if the V-model is the basis here, the standard is in reality applied in many different ways, including, e.g. distributed development across multiple organizations, iterative development and agile approaches. Independent of the development approach used, the rationale of the safety standardization is argumentation that the requirements in the standards have been appropriately addressed in the product.

In the forthcoming sections we briefly describe parts 2–8 of the standard. Part 1 contains definitions and abbreviations used in the standard. The safety analysis methods described in part 9 are only covered implicitly in this chapter as they are referenced from the activities in parts 3–6. Also, part 10 is not described here as it is an informative collection of guidelines for how parts 2–6 may be applied.

8.2 Management and Support for Functional Safety

When an organization works with functional safety, there are other processes that should be established. When looking into part 2 of the ISO 26262 standard, there are requirements to have a quality management system in place, e.g. ISO 9001 [ISO15] or ISO/TS 16949 [ISO09]; to have relevant processes, including processes for functional safety, established in the management system, the sufficient competence and experience; and the field monitoring established. Field monitoring from a functional safety perspective is in particular important for detecting potential faults in electronics and software when the vehicle is in use.

During product development, there are also requirements on assigning proper responsibilities for functional safety, to plan activities related to functional safety and to monitor that the planned activities are done accordingly.

In addition, there are requirements to have proper support according to part 8, including:

- Interfaces within distributed developments, which ensure that responsibilities are clear between different organizations that share the development work, e.g. between a vehicle manufacturer and its suppliers. It is often referred to as a Statement of Work.
- Requirements management, which ensures that requirements, in particular safety requirements, are properly managed. This includes identification of requirements, requirement traceability, and status of the requirements.
- Configuration management, which ensures that changes to an Item are controlled throughout its lifecycle. There are other standards for configuration management, e.g. ISO 10007, referenced from ISO 26262.
- Change management, which in ISO 26262 ensures that functional safety is maintained when there are changes to an Item. It is based on an analysis of proposed changes and control of those changes. Change management and configuration management typically go hand in hand with each other.
- Documentation management, which in ISO 26262 ensures that all documents are managed such that they are retrievable and contain certain formalities such as unique identification, author and approver.
- Confidence in the use of software tools, which shall be done when compliance with the ISO 26262 standard relies on correct behavior of software tools used during product development, e.g. code generators and compilers. The first step is tool classification to determine if the tool under consideration is critical and, if critical, tool qualification is done to ensure that the tool can be trusted.

These requirements mean that ISO 26262 poses requirements on the product development databases described in Chap. 3 in terms of the kind of connections and relations that should be maintained.

8.3 Concept and System Development

According to ISO 26262, product development starts with the development of a concept as described in part 3. In this phase, the vehicle level function of an Item is developed. Also, the context of the Item is described, i.e. the vehicle and other technologies such as mechanical and hydraulical components. After the concept phase, there is the system development phase according to part 4 in ISO 26262. In ISO 26262, the system only contains electronic hardware and software components, no mechanical components. The development of these other components is not covered by ISO 26262.

The first step in concept development is to define the Item to which ISO 26262 is applied. This definition of the item contains functional and non-functional requirements, use of the Item including its context, and all relevant interfaces and interactions of the Item. It is an important step as this definition is the basis for continued work.

The following step is the hazard analysis and risk assessment, which includes hazard identification and hazard classification. A hazard in ISO 26262 is a potential source of harm, i.e. a malfunction of the Item that could harm persons. Examples of hazards are no airbag deployment when intended and unintended steering column lock. These hazards are then further analyzed in relevant situations, e.g. driving in a curve with oncoming traffic is a relevant situation for the unintended locking of the steering column. The combination of a hazard and relevant driving situations that could lead to harm is called hazardous event.

During hazard classification, hazardous events are classified with an ASIL. ASIL is an ISO 26262-specific term defined as Automotive Safety Integrity Level. There are four ASILs ranging from ASIL A to ASIL D. ASIL D is assigned to hazardous events that have the highest risk that needs to be managed by ISO 26262, and ASIL A to the lowest risk. If there is no ASIL, it is assigned QM, i.e. Quality Management. The ASIL is derived from three parameters, Controllability, Exposure, and Severity. These parameters estimate the magnitude of the probability of being in a situation where a hazard could result in harm to persons (Exposure), the probability of someone being able to avoid that harm given that situation and that hazard (Controllability), and an estimate of the severity of that harm (Severity). In Table 8.1, a brief explanation of the different ASILs and examples are shown.

In addition to ASIL being a measure of risk, it also puts requirements on safety measures that need to be taken to reduce the risk to an acceptable level. The higher the ASIL, the more the safety measures are needed. Examples of safety measures are analyses, reviews, verifications and validations, safety mechanisms implemented in electronic hardware and software to detect and handle fault, and independent

Table 8.1 Brief description of different ASILs with examples; the examples are dependent on vehicle type

Risk classification	Description of risk	Examples of hazardous event
QM	The combination of probability of accident (*Controllability* and *Exposure*) and severity of harm to persons (*Severity*) given the hazard is considered acceptable. With a QM classification, there are no ISO 26262 requirements on the development	No locking of steering column when leaving the vehicle in a parked position. Not possible to open sunroof
ASIL A	A low combination of probability of accident and severity of harm to persons given the hazard occurring	No airbag deployment in a crash fulfilling airbag deployment criteria
ASIL B	...	Unintended hard acceleration of vehicle during driving
ASIL C	...	Unintended hard braking of vehicle during driving while maintaining vehicle stability
ASIL D	Highest probability of accident and severity of harm to persons given the hazard occurring	Unintended locking of steering column during driving

Table 8.2 A simplified hazard analysis and risk assessment with two separate examples

Function	Hazard	Situation	Hazardous event	ASIL	Safety goal
Steering column lock	Unintended steering column lock	Driving in curve with oncoming traffic	Driver loses control of his vehicle, entering the lane with oncoming traffic	D	Steering column lock shall not be locked during driving
Driver airbags	No deployment of driver airbags	Crash where airbag should deploy	Driver is not protected by airbags in a crash when he should be	A	Driver airbag shall deploy in crash, meeting deployment criteria

safety assessments. If there is a QM, it means that there are no requirements on safety measures specified in ISO 26262. Still, normal automotive development is needed and this includes proper quality management, review, analysis, verification and validation, and much more.

For hazardous events where there is an ASIL assigned, a Safety Goal shall be specified. A Safety Goal is a top-level safety requirement to specify how the hazardous event can be avoided. A simplified hazard analysis and risk assessment is shown in Table 8.2.

The third step is the functional safety concept where each Safety Goal with an ASIL is decomposed into a set of Functional Safety Requirements and allocated to a logical design. It is also important to provide an argumentation of why the

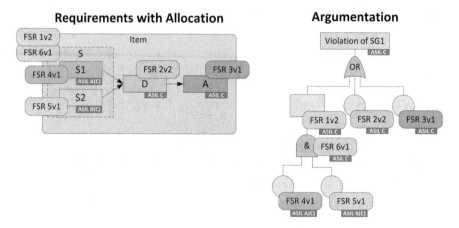

Fig. 8.2 The three parts of a functional safety concept, the Functional Safety Requirements, noted as FSR, their allocation to a logical design, and the argumentation why the Functional Safety Requirements fulfill the Safety Goal, noted as SG

Functional Safety Requirements fulfill the Safety Goal; this argumentation can be supported by a fault tree analysis.

During the Functional Safety Concept stage, and also during later refinements of safety requirements, it is possible to lower the ASILs if there is redundancy with respect to the requirements. There is always a trade-off whether to use redundancy or not. Redundant components could increase cost, at the same time as lower ASILs could save cost. The choice to make needs to be assessed on a case-by-case basis.

An example of a Functional Safety Concept is shown in Fig. 8.2. Here the logical design consists of three parts, the sensor element S, the decision element D and the actuation element A. The sensor element has been refined using redundancy of two sensor elements, S1 and S2. For all of these elements, there are Functional Safety Requirements allocated, denoted bt FSR, with a sequence number and an ASIL. For the argumentation of why these Functional Safety Requirements fulfill the Safety Goal SG1, a fault tree with the violation of the Safety Goal, SG1, as a top event is used.

During system development, as shown in Fig. 8.1 as part 4, the Functional Safety Concept is refined into a Technical Safety Concept. It is very similar to a Functional Safety Concept, but more specific in details. At this point, we work with actual systems and components, including signaling in between. It is common that a Technical Safety Concept includes interfaces, partitioning, and monitoring. The Technical Safety Concept includes Technical Safety Requirements that are allocated and an argumentation of why the Technical Safety Concept fulfills the Functional Safety Concept (Fig. 8.3). An example of one possible level of design for a Technical Safety Concept is shown in Fig. 8.4. Here the design for the decision element has been refined to an ECU that consists of a microcontroller and an ASIC. For these

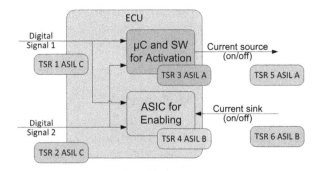

Fig. 8.3 A Technical Safety Concept is one level more detailed than a Functional Safety Concept, here with a microcontroller including software (SW) and an ASIC for ensuring correct activation

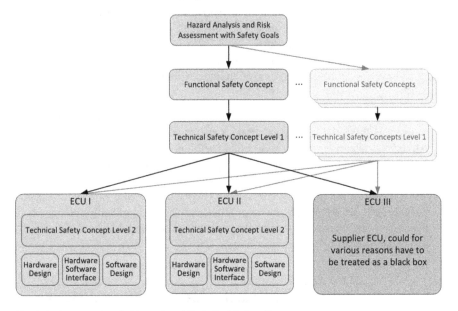

Fig. 8.4 An example of a hierarchy of Technical Safety Concepts

two elements, there are Technical Safety Requirements allocated, denoted by TSR, with a sequence number and an ASIL.

During actual development, it is common that there is a hierarchy of Technical Safety Concepts. In addition, for each Safety Goal with an ASIL there are other Functional Safety Concepts and Technical Safety Concepts. An example of the relationships between safety concepts is shown in Fig. 8.4. In this case, the top ones allocate Technical Safety Requirements to elements that consist of both software and hardware, e.g. an ECU. In the lowest one, the Technical Safety Requirements are allocated to software and hardware. In this lowest level of Technical Safety Concept, there is also a hardware-software interface. The following step in the design is detailed hardware and software development. In this chapter, we will only consider the software part. The hardware development has a similar structure as the software development.

8.4   Planning of Software Development

The software development starts with a planning phase. In addition to the planning of all software activities, including assigning resources and setting schedules, the methods and tools used need to be selected. At this phase, the modeling or programming languages to be used are also determined. The software activities to be planned are shown in Fig. 8.5 and also described in more detail in this chapter.

Even if ISO 26262 is described in a traditional context with manually written code according to a waterfall model, ISO 26262 supports both automatic code generation and agile way of working.

To support the development and to avoid common mistakes, there is a requirement to have modeling and coding guidelines. These shall address the following aspects:

- Enforcement of low complexity: ISO 26262 does not define what low complexity is and it is up to the user to set an appropriate level of what is sufficiently low. An appropriate compromise with other methods in this part of ISO 26262 may be required. One method that can be used is to measure cyclomatic complexity and have guidance for what to achieve.
- Use of language subsets: When coding, depending on the programming language, there are language constructs that may be ambiguously understood or may easily lead to mistakes. Such language constructs should be avoided, e.g. by using MISRA-C [A+08] when coding in C.

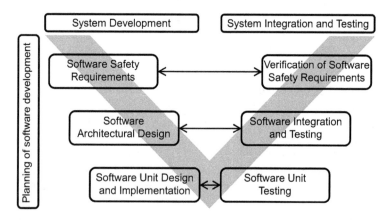

Fig. 8.5 The software development activities according to ISO 26262. Adopted from [ISO1]

- Enforcement of strong typing: Either strong typing is inherent in the programming language used, or there shall be principles added to support this in the coding guidelines. The advantage of strong typing is that the behavior of a piece of software is more understandable during design and review as the behavior has to be explicit. When strong typing is inherent in the programming language, a value has a type and what can be done with that value depends on the type of the value, e.g. it is not possible to add a number to a text string.
- Use of defensive implementation techniques: The purpose of defensive implementation is to make the code robust to continue to operate even in the presence of faults or unforeseen circumstances, e.g. by catching or preventing exceptions.
- Use of established design principles: The purpose is to reuse principles that are known to work well.
- Use of unambiguous graphical representation: When using graphical representation, it should not be open to interpretation.
- Use of style guides: A good style when coding typically makes the code maintainable, organized, readable, and understandable. Hence, the likelihood for faults is lowered when using good style guides. One example of a style guide for C [A+08] is [DV94]
- Use of naming conventions: By using the same naming conventions the code becomes easier to read, e.g. by using *Title Case* for names of functions.

8.5 Software Safety Requirements

Once we have Technical Safety Requirements allocated to software and the software development planned, it is time to specify the software safety requirements. These are derived from the Technical Safety Concept and the system design specification, while also considering the hardware-software interface. At the end of this step, we shall also verify that the software safety requirements, including the hardware-software interface, realize the Technical Safety Concept.

In a safety-critical context, there are several services expected from software that are specified by software safety requirements, including:

- Correct and safe execution of the intended functionality.
- Monitoring that the system maintain a safe state.
- Transitioning the system to a degraded state with reduced or no functionality, and keeping the system in that state.
- Fault detection and handling hardware faults, including setting diagnostic fault codes.
- Self-testing to find faults before they are activated.
- Functionality related to production, service and decommissioning, e.g. calibration and deploying airbags during decommissioning.

8.6   Software Architectural Design

The software safety requirements need to be implemented in a software architecture together with other software requirements that are not safety-related. In the software architecture, the software units shall be identified. As the software units get different software safety requirements allocated to them, it is also important to consider if these requirements, potentially with different ASILs, can coexist in the same software unit. There are certain criteria to be met for coexistence. If these criteria aren't met, the software needs to be developed and tested according to the highest ASIL of all allocated safety requirements. These criteria may include memory protection and guaranteed execution time.

The software architecture includes both static and dynamic aspects. Static aspects are related to interfaces between the software units and dynamic aspects are related to timing, e.g. execution time and order. An example can be seen in Fig. 8.6. To specify these two aspects, the notation of the software architecture to be used is informal, semi-formal or formal. The higher the ASIL, the more the formality needed.

It is also important that the software architecture consider maintainability and testability. In an automotive context, software needs to be maintainable as its lifetime is considerable. It is also necessary that the software in the software architecture easily be tested, as testing is important in ISO 26262. During the design of the software architecture, it is also possible to consider the use of configurable software. There are both advantages and disadvantages when using it.

To avoid systematic faults in software resulting from high complexity, ISO 26262 specifies a set of principles that shall be used for different parts, including:

- Components shall have a hierarchical structure, shall have high cohesion within them and be restricted in size.
- Interfaces between software units shall be kept simple and small. This can be supported by limiting the coupling between software units by separation of concerns.

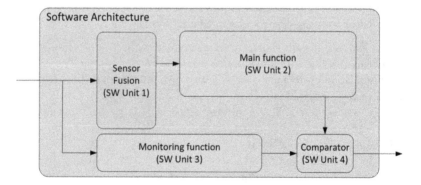

Fig. 8.6 A simple software architecture with four software units

- Scheduling of software units shall be appropriate based on the software, and if interrupts are used, these shall be avoided and be priority-based. The purpose is to ensure timely execution of software units.

At the software architectural level there is a good possibility of detecting errors between different software units. As in general for different ASILs, the higher the ASIL, the more the mechanisms needed. These are mechanisms mentioned in ISO 26262, some overlapping with each other:

- Range checks of data: This is a simple method to ensure that the data read from or written to an interface is within a specified range of values. Any value outside this range is to be treated as faulty, e.g. a temperature below absolute zero.
- Plausibility checks: This is a type of sanity check that can be used on signals between software units. It should, e.g., catch a vehicle speed signal going from standstill to 100 km/h in 1 s for a normal car. Such acceleration is not plausible. A plausibility check could use a reference model or compare information from other sources to detect faulty signals.
- Detection of data errors: There are many different ways of detecting data errors, e.g. error detecting codes such as checksums and redundant data storage.
- External monitoring facility: To detect faults in execution, an external monitoring facility can be quite effective. It can, e.g., be software executed in a different microcontroller or a watchdog.
- Control flow monitoring: By monitoring the execution flow of a software unit, certain faults can be detected, including skipped instructions and software stuck in infinite loops.
- Diverse software design: Using diversity in software design can be efficient. The approach is to design two different software units monitoring each other; if the behaviors differ, there is a fault that should be handled. This method can be questioned, as it is not uncommon that software designers make similar mistakes. To avoid similar mistakes, the more diverse the software functionality is, the lower the likelihood of these types of mistakes.

Once an error has been detected, it should be handled. The mechanisms for error handling at the software architectural level specified in ISO 26262 are:

- Error recovery mechanism: The purpose is to go from a corrupted state back into a state from which normal operation can be continued.
- Graceful degradation: This method takes the system from a normal operation to a safe operation when faults are detected. A common example in automotive software is to warn the driver that something is not working by a warning lamp, e.g. the airbag warning lamp when the airbags are unavailable.
- Independent parallel redundancy: This type of mechanism may be quite costly as it may need redundant hardware to be efficient. The concept is based on the assumption that the likelihood of simultaneous failures is low and one redundant channel should always be operating safely.
- Correcting codes for data: For data errors, there are mechanisms that can correct these. These mechanisms are all based on adding redundant data to give different

levels of protection. The more the redundant data that is used, the more the errors that can be corrected. This is typically used on CDs, DVDs, and RAM, but can be used in this area as well.

Once the software architectural design is done, it needs to be verified against the software requirements. ISO 26262 specifies a set of methods that are to be used:

- Walk-through of the design: This method is a form of peer review where the software architecture designer describes the architecture to a team of reviewers with the purpose to detect any potential problems.
- Inspection of the design: In contrast to a walk-through, an inspection is more formal. It consists of several steps, including planning, off-line inspection, inspection meeting, rework and follow-up of the changes.
- Simulation: If the software architecture can be simulated, it is an effective method, in particular for finding faults in the dynamic parts of the architecture.
- Prototype testing: As with simulation, prototyping can be quite efficient for the dynamic parts. It is however important to analyze any differences between the prototype and the intended target.
- Formal verification: This is a method, rarely used in the automotive industry, to prove or disprove correctness using mathematics. It can be used to ensure expected behavior, exclude unintended behavior, and prove safety requirements.
- Control flow analysis: This type of analysis can be done during a static code analysis. The purpose is to find any safety-critical paths in the execution of the software at an architectural level.
- Data flow analysis: This type of analysis can also be done during a static code analysis. The purpose is to find safety-critical values of variables in the software at an architectural level.

8.7 Software Unit Design and Implementation

Once the software safety requirements are specified and the software architecture down to software unit level is ready, it is time to design and implement the software units. ISO 26262 supports both manually written code and automatically generated code. If the code is generated, the requirements on software units could be omitted, given that the tool used can be trusted, as determined by tool classification, and if needed tool qualification. In this section, the focus will be on manually written code.

As with the specification of the software architecture, ISO 26262 specifies the notation that should be used for the software unit design. ISO 26262 requires an appropriate combination of notations to be used. Natural language is always highly recommended. In addition the standard recommends informal notation, semi-formal notation and formal notation. Formal notation is not really required at this time.

There are many design principles mentioned in ISO 26262 for software unit implementation. Some may not be applicable, depending on the development. Many could also be covered by the coding guidelines used. However, all are mentioned here for completeness:

- One entry and one exit point: One main reason for this rule is to have understandable code. Multiple exit points complicate the control flow through the code and therefore the code is harder to understand and to maintain.
- No dynamic objects or variables: There are two main challenges with dynamic objects and variables, unpredictable behavior and memory leaks. Both may have a negative effect on safety.
- Initialization of variables: Without initializing variables, anything can be put in them, including unsafe and illegal values. Both of these may have a negative effect on safety.
- No multiple use of variable names: Having different variables using the same name risks, confusing to readers of the code.
- Avoid global variables: Global variables are bad from two aspects; they can be read by anyone and be written to by anyone. Working with safety-related code, it is highly recommended to have control of variables from both aspects. However, there may be cases where global variables are preferred, and ISO 26262 allows for these cases if the use can be justified in relation to the associated risks.
- Limited use of pointers: Two significant risks of using pointers are corruption of variable values and crashes of programs; both should be avoided.
- No implicit type conversions: Even if supported by compilers for some programming languages, this should be avoided as it could result in unintended behavior, including loss of data.
- No hidden data flow or control flow: Hidden flows make the code harder both to understand and to maintain.
- No unconditional jumps: Unconditional jumps make the code harder to analyze and understand with limited added benefit.
- No recursions: Recursion is a powerful method. However, it complicates the code, making it harder to understand and to verify.

At the time of software unit design and implementation, it is required to verify that both the hardware-software interface and the software safety requirements are met. In addition, it shall be ensured that the implementation fulfills the coding guidelines and that the software unit design is compatible with the intended hardware. To achieve this, there are several methods to be used:

- Walk-through.[1]
- Inspection (see footnote 1).
- Semi-formal verification: This family of methods is between informal verification, like reviews, and formal verification, with respect to ease of use and strength in verification results.
- Formal verification (see footnote 1).

---

[1]See Sect. 8.6.

- Control flow analysis (see footnote 1).
- Data flow analysis (see footnote 1).
- Static code analysis: The basis for this analysis is to debug source code without executing it. There are many tools with increasing capabilities. These often include analysis of syntax and semantics, checking coding guidelines like MISRA-C [A+08], variable estimation, and analysis of control and data flows.
- Semantic code analysis: This is a type of static code analysis considering the semantic aspects of the source code. Examples of what can be detected include variables and functions not properly defined and used in incorrect ways.

8.8 Software Unit Testing

The purpose of testing the implemented and verified software units is to demonstrate that the software units meet their software safety requirements and do not contain any undesired behavior, as shown in Fig. 8.7. There are three steps needed to achieve this purpose; an appropriate combination of test methods shall be used, the test cases shall be determined, and an argumentation of why the test done gives sufficient coverage shall be provided. It is also important that the test environment used for the software unit testing represent the target environment as closely as possible, e.g. model-in-the-loop tests and hardware-in-the-loop tests as described in Chap. 3.

ISO 26262 provides a set of test methods, and an appropriate combination shall be used, depending on the ASIL of the applicable software safety requirements. The test methods for software unit testing are:

- Requirements-based test: This testing method targets ensuring that the software under test meets the applicable requirements.
- Interface test: This testing method target to ensure that all interactions with the software under test work as intended. It should also detect any incorrect

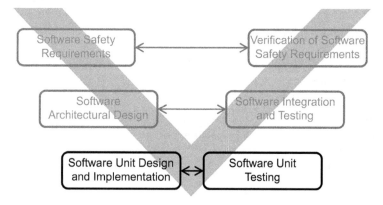

Fig. 8.7 Software unit testing is done at the level of software unit design and implementation

assumptions made on the interfaces under test. These interactions should have been specified by requirements and hence this testing method is overlapping with requirements-based tests.

- Fault injection test: This method is a very efficient test method for safety-related testing. The key part is to test to see if there is something missing in the test target. By injecting different types of faults, together with monitoring and analyzing the behavior, it is possible to find weaknesses that need to be fixed, e.g. by adding new safety mechanisms.
- Resource usage test: The purpose of this test method is to verify that the resources, e.g. communication bandwidth, computational power and memory, are sufficient for safe operation. For this type of testing, the test target is very important.
- Back-to-back comparison test: This method compares the behavior of a model with the behavior of the implemented software when both are stimulated in the same way. Any differences in behavior could be potential faults that need to be addressed.

Similarly, ISO 26262 provides a set of methods for deriving test cases for software unit testing. These methods are:

- Analysis of requirements: This method is the most common approach for deriving test cases. Basically, the requirements are analyzed and a set of appropriate test cases are specified.
- Generation and analysis of equivalence classes: The purpose of this method is to reduce the number of test cases needed to give good test coverage. This is done by identifying equivalence classes of input and output data that test the same condition. Test cases are then specified with the target to give appropriate coverage.
- Analysis of boundary values: This method complements equivalence classes. The test cases are selected to stimulate boundary values of the input data. It is recommended to consider the boundary values themselves, values approaching and crossing the boundaries and out of range values.
- Error guessing: The advantage of this method is that the test cases are generated based on experience and previous lessons learned.

The last step in software unit testing is to analyze if the test cases performed provide sufficient test coverage. If this isn't the case, more tests need to be carried out. The analysis of coverage is according to ISO 26262, done using these three metrics:

- Statement coverage: The goal is to have all statements, e.g. *printf ("Hello World! n")*, in the software executed.
- Branch coverage: The goal is to have all branches from each decision statement in the software executed, e.g. both true and false branches from an if statement.
- Modified Condition/Decision Coverage (MC/DC): The goal of this test coverage is to have four different criteria met. These are: each entry and exit point is executed, each decision executes every possible outcome, each condition in a

decision executes every possible outcome, and each condition in a decision is shown to independently affect the outcome of the decision.

8.9   Software Integration and Testing

Once all software units have been implemented, verified and tested, it is time to integrate the software units and to test the integrated software. At this testing, the target is to test the integrated software against the software architectural design, as shown in Fig. 8.8. This testing is very similar to software unit testing and consists of three steps: selection of test methods, specification of test cases, and analysis of test coverage. Also, the test environment shall be as representative as possible.

The test methods for software integration testing are the same as for software unit testing as described in Sect. 8.8, namely:

- Requirements-based test
- Interface test
- Fault injection test
- Resource usage test
- Back-to-back comparison test

The methods for deriving test cases for software integration testing are the same as for software unit testing as described in Sect. 8.8, namely:

- Analysis of requirements
- Generation and analysis of equivalence classes
- Analysis of boundary values
- Error guessing

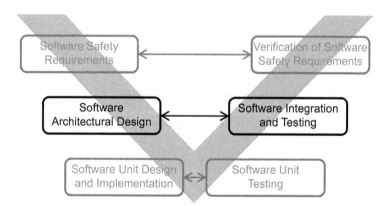

Fig. 8.8 Software unit integration and testing is done at the level of software architecture

The last step in the testing of the integrated software is to analyze test coverage. Again, if the coverage is too low, more tests need to be done. The analysis of coverage according to ISO 26262 is done using the following methods:

- Function coverage: The goal of this method is to execute all functions in the software.
- Call coverage: The goal of this method is to execute all function calls in the software. The key difference of this coverage compared with function coverage is that a function may be called from many different places and ideally all of these calls are executed during testing.

8.10 Verification of Software Safety Requirements

Once the software has been fully integrated, it is time for verification of the software against the software safety requirements, as shown in Fig. 8.9. ISO 26262 specifies possible test environments that can be used. At this point in time, the environment to use is very dependent on the type of development. These test environments may include a combination of:

- Hardware-in-the-loop: Using actual target hardware in combination with a virtual vehicle is a cost-efficient way of testing. As it uses a virtual vehicle, it should be complemented by another environment.
- Electronic control unit network environments: Using actual hardware and software for the external environment is quite common. It is more correct compared to a virtual vehicle; at the same time it may be less efficient in running the tests.
- Vehicles: Using vehicles during this level of testing is in particular useful when there is software that is in operation and has been modified. At the same time, it makes for the most costly test environment.

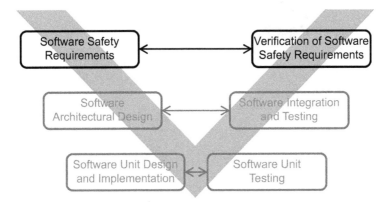

Fig. 8.9 Software unit integration and testing is done at the level of software architecture

8.11   Examples of Software Design

In this section we take some brief examples from the previous sections to show how ISO 26262 could impact software design. In the example in Fig. 8.10, we have an assumed Safety Goal covering faulty behavior classified as ASIL D, and no other Safety Goal. This example has also broken down the ASIL D into two independent ASIL B channels using ASIL decomposition. However, the comparator in the end needs to meet ASIL D requirements as it is a single point of failure.

From the early phases of planning, there will be a requirement on the programming language used, as shown in Fig. 8.10; when using the C language, the MISRA C [A+08] standard is common. An example of a software safety requirement for the comparator in the figure is to transition to a safe state in case of detected errors. In this example a safe state could be no functionality, a so-called fail-silent state. As intentionally shown in Fig. 8.10, working with the software architectural design is quite important. In this example we see the plausibility and range checks on the sensor side as well as external monitoring using diverse software. To make full benefit of this monitoring function, it needs to be allocated to independent hardware. For the testing of the main function, using methods meeting ASIL B requirements on testing is sufficient.

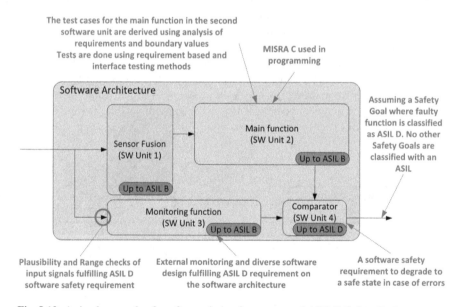

Fig. 8.10 A simple example of a software design for an assumed ASIL D Safety Goal

8.12 Integration, Testing, Validation, Assessment and Release

Once we have fulfilled the Technical Safety Requirements in the design and implementation of software and hardware and also shown by testing that the derived requirements are fulfilled, it is time to integrate hardware and software. In ISO 26262, this is done at three different levels; hardware-software, system and vehicle. At each level, both integration and testing are required. In real development, there can be fewer integration levels and more integration levels, especially when the development has been distributed among vehicle manufacturers and suppliers at many different levels. At each level, there are specific methods to derive test cases and methods to be used during testing. All of these have the purpose of providing evidence that the integrated elements work as specified.

Once we have our Item integrated in a vehicle we can finalize the safety validation. The purpose of safety validation is to provide evidence that the safety goals and the functional safety concept are appropriate and achieved for the Item. By doing so, we have finalized the development and the only remaining activities are to assess and to conclude that the development has resulted in a safe product.

To document the conclusion and the argument that safety has been achieved, a safety case is written. A safety case consists of this argumentation, with pointers to different documents as evidence. Typical evidence includes hazard analysis and risk assessment, safety concepts, safety requirements, review reports, analysis reports, and test reports. It is recommended that the safety case be written in parallel with product development, even if it can't be finalized before the development activities have been finalized.

Once the safety case has been written, it is time for functional safety assessment of Items with higher ASILs. There are many details on how this is to be done, but let us simplify it here. Basically, someone independent is to review the developed system, the documentation that led to the system, in particular the safety case, and the ways of working during the development. If the person doing the assessment is satisfied, it is possible to do the release for production and start producing.

8.13 Production and Operation

Functional safety as a discipline mainly focuses on product development. At the same time, what is developed needs to be produced and is intended to be used in operation by users of the vehicle. Part 7 of ISO 26262 is the smallest part of the whole standard and describes what is required during both production and operation. In addition, planning for both production and operation are activities to be done in parallel to product development.

The requirements for production can be summarized so as to produce what was intended, including maintenance of a stable production process, documentation of what was done during production if traceability is necessary, and carrying out of needed activities such as end-of-line testing and calibration.

For operation, there are clear requirements on information that the driver and service personnel should be aware of in, e.g., instructions in a driver's manual, service instructions and disassembly instructions. One key part during operation is also a field monitoring process. The purpose of this process is to detect potential faults, analyze those faults, and, if needed, initiate proper activities for vehicles in operation.

8.14   Further Reading

In this chapter, we have provided an overview of ISO 26262 and gone into details of the software-specific parts. For more details on both of these parts, the ISO 26262 standard itself [ISO11] is a good alternative when starting to work, especially for the software-specific parts. At the same time, understanding this standard, like many standards, would benefit from basic training to get the bigger picture and the logic behind it. For more details on safety-related software, the work in [HHK10] gives a good start.

To go into details of functional safety in general, there are some books available. One of the classical books that gives a good overview, even if it is a bit old, is [Sto96]. A newer book has been written by Smith [SS10]. This book gives a good overview of functional safety standards in general and details of the IEC 61508 and IEC 61511 standards. Even if these are different from ISO 26262, the book still gives good insights that can be used in an automotive context.

When working with functional safety, it is apparent that much of the work is based on various safety analyses. There is one book that gives a good overview of what is most used in an automotive context, written by Ericson [E+15]. This book is well worth reading.

Also, one of the key parts in ISO 26262 and many other safety standards is the argumentation for safety. This is often documented in a Safety Case. To understand more on Safety Cases, [WKM97] gives a good overview. For the argumentation part, the Goal Structuring Notation is both well recognized and an effective approach. This is well described in [KW04].

8.15   Summary

In this chapter we have described how the automotive industry works with functional safety and in particular focused on software development. As apparent in this section, the ISO 26262 standard is the basis for this in the automotive industry. It is quite a significant standard and it is more or less a prerequisite for being in the industry, both for organizations and for individuals.

It is not a standard that is possible to learn overnight, at the same time it is fairly straightforward for some parts like software engineering. As seen in this section, the software-specific details in ISO 26262 is more or less a set of additional rules that one adheres to following normal software development practices.

The reader should also have seen what is typical of ISO 26262; there is no single answer. This is a standard that describes a simplified way of working with functional safety in the automotive industry. As there are many different types of development, this standard has to be adapted to fit each type of development. Hence, the user of this standard has both a lot of flexibility when applying it and at the same time a lot of responsibility with regard to arguing for the choices made, e.g. for the test methods chosen when testing a software unit. There are also differences in how the standard is interpreted in, e.g., different nations, types of vehicles and levels in the supply chain.

There is currently an ongoing revision of ISO 26262 [ISO16], where the significant changes will be in the scope of the standard, now to include all road vehicles except mopeds. There will also be two new parts, one informative part for semiconductors and one normative part for motorcycles. In addition, from the ISO 26262 community there has been work initiated to cover safety of fault-free vehicle-level functions as well as automotive security. These are not yet ready ISO standards.

References

A⁺08. Motor Industry Software Reliability Association et al. *MISRA-C: 2004: guidelines for the use of the C language in critical systems*. MIRA, 2008.

DV94. Jerry Doland and Jon Valett. C style guide. *NASA*, 1994.

E⁺15. Clifton A Ericson et al. *Hazard analysis techniques for system safety*. John Wiley & Sons, 2015.

HHK10. Ibrahim Habli, Richard Hawkins, and Tim Kelly. Software safety: relating software assurance and software integrity. *International Journal of Critical Computer-Based Systems*, 1(4):364–383, 2010.

IEC10. IEC. 61508:2010 – functional safety of electrical/electronic/programmable electronic safety-related systems. *Geneve, Switzerland*, 2010.

ISO09. ISO. Quality management systems – particular requirements for the application of iso 9001:2008 for automotive production and relevant service part organizations. *International Standard ISO/TS*, 16949, 2009.

ISO11. ISO. 26262–road vehicles-functional safety. *International Standard ISO*, 26262, 2011.

ISO15. ISO. 9001: 2015 quality management system–requirements. *Geneve, Switzerland*, 2015.

ISO16. ISO. 26262–road vehicles-functional safety. *International Standard ISO*, 26262, 2016.

KW04. Tim Kelly and Rob Weaver. The goal structuring notation–a safety argument notation. In *Proceedings of the dependable systems and networks 2004 workshop on assurance cases*. Citeseer, 2004.

SS10.   David J Smith and Kenneth GL Simpson. *Safety Critical Systems Handbook: A Straightfoward Guide To Functional Safety, IEC 61508 (2010 Edition) And Related Standards, Including Process IEC 61511 And Machinery IEC 62061 And ISO 13849.* Elsevier, 2010.

Sto96.   Neil R Storey. *Safety critical computer systems.* Addison-Wesley Longman Publishing Co., Inc., 1996.

WKM97. SP Wilson, Tim P Kelly, and John A McDermid. Safety case development: Current practice, future prospects. In *Safety and Reliability of Software Based Systems*, pages 135–156. Springer, 1997.

Chapter 9
Current Trends in Automotive Software Architectures

Abstract Cars have evolved a lot since their introduction and will evolve even more. Today's cars would not work without the software that is embedded in their electronics. Although the physical processes are often the same as in the cars' of the 1990s (combustion engines, servo steering), they become computer platforms and are able to "think" and drive autonomously. In this chapter we look into a few trends which shape automotive software engineering—autonomous driving, self-* systems, big data and new software engineering paradigms. We look into how these trends can shape the future of automotive software engineering.

9.1 Introduction

Automotive software evolves over time and requires changes to the methods used to develop it. The evolution of software means that we can use new functions which require more software, but also that we can use more advanced software development methods.

If we look at the history of electronics and software in cars, we can see that it is today that the big technological breakthroughs are happening. The cars of today have become sophisticated computer platforms which can be used in multiple ways. The powertrain technology has changed from traditional combustion engines to electrical or hybrid (e.g. hydrogen technology).

Living in these interesting times, software engineers and architects will see a lot of great possibilities and great potential. Let us then explore a few trends that seem to shape current automotive software engineering. In particular, let us explore the following trends:

- Autonomous driving—how the introduction of autonomous driving shapes the automotive sector and the software needed to steer cars.
- Self-*—how the ability to develop self-healing and self-adaptive systems influences the way in which we can design software in modern cars.
- Big data—how the ability to communicate and process large quantities of data changes the way we think about decision making in cars.
- New software development paradigms—how new software engineering methods influence the way we develop software for automotive systems.

In the remainder of this chapter we go through these trends.

© Springer International Publishing AG 2017
M. Staron, *Automotive Software Architectures*,
DOI 10.1007/978-3-319-58610-6_9

9.2 Autonomous Driving

Undoubtedly the main trend in modern software in cars' is autonomous driving software. Autonomous driving software allows drivers to skip controlling the car or some of its functions. The NHSTA (National Highway Safety Traffic Administration) in the United States recognizes the following levels of autonomous functionality in cars [A+13]:

- Level 0, No automation—there are no functions in the car that can drive the car or support the driver.
- Level 1, Function-specific automation—according to the definition "automation at this level involves one or more specific control functions", meaning that certain functions can be autonomous, e.g. adaptive cruise control.
- Level 2, Combined function automation—where a group of functions can be automated and be autonomous. The driver, however, is still responsible for the control of the vehicle and must be prepared to take control of the vehicle on very short notice. Example functions are self-driving on highways.
- Level 3, Limited self-driving automation—the vehicle is able to drive autonomously under certain conditions and monitor the conditions; the drivers might need to occasionally take control, but the transition time is comfortably longer than at level 2.
- Level 4, Full self-driving automation—the vehicle is able to perform the entire trip autonomously; the driver is only expected to enter constraints and the destination for the trip. The level applies to both manned and unmanned vehicles.

One can see that modern vehicles already provide functions for automation Level 2 (combined function automation) and some even for level 3 (e.g. Tesla's autopilot functionality, [Pas14, Kes15]). This kind of functionality puts a lot of constraints on the automotive software.

First of all, this drives the complexity of software and therefore the cost of its development, verification, validation and certification. As the self-driving functionality is safety-critical it requires specific validation. It also requires complex reasoning in traffic situations on a very abstract level—e.g. whether it is better to save lives of the car's passengers or the lives of others in the accident.

Second of all, this kind of functionality drives the need for large quantities of data to process, which drives the need for processing power in modern cars. The processing power requires efficient CPUs and electronic buses of high throughput, which require more advanced infrastructure (e.g. cooling fans), that is often susceptible to environmental influences such as vibrations, humidity and temperature. This means that new components need to be develop especially for the cars, which drives costs.

Third of all, we need to understand that the quality of the sensors today is insufficient for advanced scenarios. Cameras are able to see clearly in specific conditions, but the human eye is still better synchronized with the human brain in all situations. Therefore cameras are not able to work effectively in low light or bad weather conditions [KTI$^+$05]. Using high-end cameras and sophisticated equipment would drive up the cost and still not guarantee the same quality as from human eyes and brains.

And finally, this kind of autonomous functionality requires acting on higher abstraction levels. Information about distance to the nearest obstacle needs to be transformed to a worldview which can be compared to a map view to determine the best course of action in a specific situation [BT16]. This requires more advanced algorithms which can be based on heuristics. The heuristics, however, are very challenging to prove to work correctly in all kinds of traffic situations, thus posing problems for safety certification.

9.3 Self-*

Self-healing is the ability of the system to autonomously change its structure so that its behaviour stays the same. An example concept of self-healing can be seen in the work of Keromytis et al. [Ker07], who define the self-healing as the ability to autonomously recover from erroneous execution.

One of the most prominent mechanisms used in self-healing systems is the MAPE-K (Measure, Analyse, Plan and Execute + Knowledge, [MNSKS05]). It is shown in Fig. 9.1 as an overwatch algorithm for an ECU realizing the adaptive cruise control functionality.

The algorithm in short is based on monitoring the execution of the algorithm for correctness. In the example of adaptive cruise control, we can monitor the radar to confirm it provides reliable results (e.g. no distortion is present). The analysis component checks whether one of the failure conditions has been detected (e.g. too

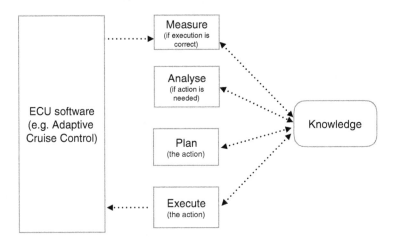

Fig. 9.1 Realization of MAPE-K for ECU software

much noise in the radar readings) and sends a signal to the plan component which plans appropriate action based on the reading and analysis. One of the actions can be to disable the adaptive cruise control and inform the user. Once the component makes a decision about the recovery strategy it moves to the execution and executes the repair strategy (i.e. informs the user and disables the adaptive cruise control algorithm).

This trend of using self-adaptation is used increasingly in safety-critical systems as it allows us to change the operation of a component in the presence of errors and failures. It can provide the ability to the system to self-degrade the functionality (e.g. temporarily change the operation of the engine, as discussed in Chap. 6).

However, there are still challenges which need to be addressed in order to make self-adaptation even more applicable to automotive systems. One of the major challenges is the ability to prove that the system is "safe" (in the sense of ISO/IEC 26262) during self-adaptation. Another is the fact that self-adaptation algorithms can be complex and need to be validated, but in many situations the failure modes cannot be replicated in real life. For example, it is difficult to safely replicate the situation where a radar in adaptive cruise control is broken when a vehicle drives at 150 km/h.

Nevertheless, we can perceive more self-* algorithms entering automotive systems as they need to monitor the increasingly complex decision algorithms in modern cars (e.g. related to autonomous driving).

9.4 Big Data

With the ability of modern cars to communicate with each other and the ability to use their own sensors in decision making, the amount to data used in modern cars has increased exponentially. At the same time, the field of computer science has evolved and started to tackle challenges related to storing, analysing and processing large quantities of data [MCB$^+$11, MSC13].

Big Data systems are often characterized by the so-called five Vs:

- Volume—big data systems have large amounts of data (e.g. tera- or petabytes), which makes storage and processing a challenging task requiring new types of algorithms.
- Variety—the data comes from heterogeneous sources, has different formats, and has multiple semantic models, which require preprocessing before the data can be fed to analysis algorithms.
- Velocity—the data is provided at high speeds and requires processing realtime (e.g. from multiple sensors in the car and needs to be used to make safety-critical decisions). The speed requires large processing power, which might not be available in such systems as the automotive software.
- Value—the data collected has some business value (e.g. data about the driving routines of cars) which makes the storage, privacy and security issues

challenging, especially in combination with the velocity of processing and the next V—veracity.

- Veracity—the data has varying degree of quality, e.g., in terms of accuracy and trustworthiness. This varying degree of accuracy makes it challenging for the systems to use.

The challenges of using big data in automotive systems are related to all of the above V's. The large volume of data which comes from the car's own sensors needs to be processed and often stored, which puts requirements on storage in cars. Before the popularization of the SSD (Solid State Disk) technology it was rather challenging to use hard disks to store data (durability problems due to vibrations). Now, it is possible to store more data and also to process more data.

The high speed of processing requires more processing power, more efficient processors which take power and more connectivity. This drives the cost of automotive hardware since the more efficient processors require more infrastructure (stability, cooling), which is prone to problems in the automotive environment (humidity, vibrations). The hardware price is so important in the automotive domain (as opposed to other domains, where hardware is considered cheap) that one usually takes a calculation (a rule of thumb) that one dollar more expensive hardware per ECU can lead to 100 dollars more expensive cars.

The veracity of the data is a challenge as in many cases the "true" values cannot be measured but computed. For example, the slippage of the road in winter conditions cannot be measured but are derived either from ABS usage or the steering wheel friction. In some cases the data is obfuscated in order to secure privacy (e.g. triangulation algorithms to hide the true position of a car), which prevent the algorithms from "knowing" the true value of the data point [SS16].

In the future we will see more of big data, as large quantities of data are needed for autonomous driving and for advanced algorithms for collision prevention and avoidance.

9.5   New Software Development Paradigms

Software engineering for automotive systems has evolved the pace of the automotive domain. So, let us look into a few of the trends which shape the field today and will potentially shape the field in the future.

Agility in Specification Development   Agile software development has been used in many domains outside of the automotive and now there is evidence that it is used increasingly in the automotive domain. In particular, at the lower part of the V-model suppliers work more agilely with their requirements engineering and software development [MS04]. We can also observe these trends scaling up to complete vehicle development [EHLB14] and [MMSB15]. With this increased adoption of Agile principles we can foresee the increased ability to specify requirements alongside software development, especially as the trends in automotive electronics increasingly contain more commodity (or off-the-shelf) components. AUTOSAR also prescribes a standardized approach to development, which eases the use of iterative development principles as the development of electronics/hardware is decoupled from the development of functions/software.

Increased Focus on Traceability The increased amount of software in cars and their increased presence in safety systems leads to stricter processes for keeping track of requirements for safety-critical systems. ISO 26262 (Road vehicles—Functional Safety) is one example of this. In the automotive domain this means that the increased complexity of software modules [SRH15] leads to more fine-grained traceability management. One of the enablers of this increased traceability is the increased integration between the tools—tool chaining [BDT10] and [ABB+12].

Increased Focus on Non-functional Properties The increased use of software for active safety systems calls for increased focus on non-functional properties of software. The increased traffic on communication buses within the car, and the increased capacity of the communication buses call for more synchronization and verification. Safety analyses such as control path monitoring, safety bits and data complexity control, are just a few examples [Sin11]. As the focus of requirements engineering research in the automotive domain was mainly (or implicitly) in the functional requirements, we foresee an increased growth of research and emphasis on the non-functional requirements.

Increased Focus on Security Requirements A dedicated group of requirements is the security requirements, as our cars are increasingly connected and therefore prone to hacker attacks [SLS+13] and [Wri11]. The recent demonstration of the possibility of steering a Jeep Wrangler vehicle offroad showed that the threat is real and related to the safety of cars and transport systems. We therefore perceive that the ability to prevent attacks will the focus of the automotive software development increasingly more in the coming decade.

9.5.1 Architecting in the Age of Agile Software Development

Architecture development in software development is usually conducted by experienced architects, and the larger the product, the more the experience required. As each type of system has its specific requirements, the architectural design requires attention to specific aspects like realtime properties or extensibility. For example, in the telecom domain the extensibility and performance are the main aspects, whereas in the automotive domain it is safety and performance that are of the utmost priority. The architecture development efforts are dependent to some extent on the software development process adopted by the company, e.g. the architecture development methods differ in the V-model and Agile methodologies. In the V-model the architecture work is mostly prescriptive and centralized around the

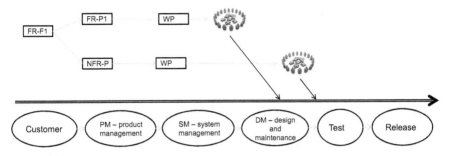

Fig. 9.2 Feature development in Lean/Agile methods

architects whereas in the Agile methods the work can be more descriptive and distributed into multiple self-organized teams.

As Agile software development principles spread in industry, architecture development evolved. As Agile development teams became self-organized, architecture work became more distributed and harder to control centrally [Ric11]. The difficulties stem from the fact that Agile teams value independence and creativity [SBB+09] whereas architecture development requires stability, control, transparency and proactivity [PW92]. Figure 9.2 presents an overview of how the functional requirements (FR) and non-functional requirements (NFR) are packaged into work packages and developed as features by teams. Each team delivers code to the main branch. Each team has the possibility to deliver the code to any component of the product.

The requirements come from the customers and are prioritized and packaged into features by product management (PM), which communicates with system management (SM) on the technical aspects of how the features affect the architecture of the product. System management communicates with the teams (DM, Test) that design, implement and test (functional testing) the feature before delivering it to the main branch. The code in the main branch is tested thoroughly by dedicated test units before being release [SM11].

9.6   Other Trends

Bosch has presented three trends which shaped software engineering in the mid-2010s [Bos16]: speed of software development, ecosystems and data-driven development. He predicted that the companies which are the first ones on the market would be more successful than others as the innovation model is based on the shark's tail rather than the traditional technology adoption curve. In particular, the majority of new, innovative software products are adopted by the market at a tremendous pace, and then companies need to be prepared to be ready for the market. Followers do not have the same ability to attract customers [DN14]. Ecosystem thinking (e.g.

Apple's App store or Google's Play store) has been present in the automotive sector from way back in the hardware domain (e.g. customers of BMW are bound to buy spare parts from manufacturer) but not in the software domain. And finally we have data-driven development and the Lean innovation thinking [Rie11] where customers provide the companies with the data on how to develop their products. With connected cars and the ability to update the car software over the air we will probably see more data-driven development in the automotive industry in the coming decade.

Burton and Willis from Gartner identified five mega-trends which have the potential of shaping software engineering in the coming decades [BW15]. These mega-trends are:

- Digital Business Moves Toward the Peak of Inflated Expectations
- IoT, Mobility and Smart Machines Rapidly Approach the Peak
- Digital Marketing and Digital Workplace Quickly Move Up
- Analytics Are at the Peak
- Big Data and Cloud Make Big Moves Toward the Trough of Disillusionment

In short, these trends will drive the need for more advanced functionality of cars and the use of big data for decision making and even the development of the cars (finding out the requirements from the data rather than focus group interviews). However, they predict that the era of wearables (e.g. smartwatches) will reach the so-called "pit of disillusion" where they will probably reach the state where no more development is of interest to the customers.

In their 2016 report, Gartner Associates provide even more focus on Artificial Intelligence, Machine Learning and autonomy. We perceive these technologies as new hype in automotive software engineering, especially when combined with different levels of autonomy and self-adaptation algorithms. This will mean even more complexity and software in future cars.

9.7 Summary

To conclude this chapter let us make a speculation that future cars will be more like computer platforms where different third party companies can build applications. We can see the self-driving car of Google as an example of such a move [Gom16].

The telecommunication domain has evolved from proprietary solutions in mobile phones of the 1990s to standardized platforms and ecosystems of the smartphones of the 2010s—Android and iOS leading the field in this direction. Customers buying a new mobile phone buy a device which they can load with apps of their own choice— some free and some paid. We can see that the ability to update car's software will lead to similar trends (already visible in the infotainment domain.)

These possibilities of opening up for third party software in cars is expected to change the face of the automotive industry in the future. Commoditizing platforms and portability between vendors on the application level can cause cars to become much safer and much more fun. We can expect the cars to become hubs for all kinds of devices and integrated with wearables to provide drivers and passengers with an even better driving experience than today's. We need to live and see what the future of software in cars will bring.

References

A⁺13.   National Highway Traffic Safety Administration et al. Preliminary statement of policy concerning automated vehicles. *Washington, DC*, pages 1–14, 2013.

ABB⁺12.   Eric Armengaud, Matthias Biehl, Quentin Bourrouilh, Michael Breunig, Stefan Farfeleder, Christian Hein, Markus Oertel, Alfred Wallner, and Markus Zoier. Integrated tool chain for improving traceability during the development of automotive systems. In *Proceedings of the 2012 Embedded Real Time Software and Systems Conference*, 2012.

BDT10.   Matthias Biehl, Chen DeJiu, and Martin Törngren. Integrating safety analysis into the model-based development toolchain of automotive embedded systems. In *ACM Sigplan Notices*, volume 45, pages 125–132. ACM, 2010.

Bos16.   Jan Bosch. Speed, data, and ecosystems: The future of software engineering. *IEEE Software*, 33(1):82–88, 2016.

BT16.   Sagar Behere and Martin Törngren. A functional reference architecture for autonomous driving. *Information and Software Technology*, 73:136–150, 2016.

BW15.   Betsy Burton and David A Willis. Gartner's Hype Cycles for 2015: Five Megatrends Shift the Computing Landscape. *Recuperado de:* https://www.gartner.com/doc/3111522/gartners--hype--cycles--megatrends--shift, 2015.

DN14.   Larry Downes and Paul Nunes. *Big Bang Disruption: Strategy in the Age of Devastating Innovation*. Penguin, 2014.

EHLB14.   Ulf Eliasson, Rogardt Heldal, Jonn Lantz, and Christian Berger. Agile model-driven engineering in mechatronic systems-an industrial case study. In *Model-Driven Engineering Languages and Systems*, pages 433–449. Springer, 2014.

Gom16.   Lee Gomes. When will Google's self-driving car really be ready? It depends on where you live and what you mean by "ready" [News]. *IEEE Spectrum*, 53(5):13–14, 2016.

Ker07.   Angelos D Keromytis. Characterizing self-healing software systems. In *Proceedings of the 4th international conference on mathematical methods, models and architectures for computer networks security (MMM-ACNS)*, 2007.

Kes15.   Aaron M Kessler. Elon Musk Says Self-Driving Tesla Cars Will Be in the US by Summer. *The New York Times*, page B1, 2015.

KTI⁺05.   Hiroyuki Kurihata, Tomokazu Takahashi, Ichiro Ide, Yoshito Mekada, Hiroshi Murase, Yukimasa Tamatsu, and Takayuki Miyahara. Rainy weather recognition from in-vehicle camera images for driver assistance. In *IEEE Proceedings. Intelligent Vehicles Symposium, 2005.*, pages 205–210. IEEE, 2005.

MCB⁺11.   James Manyika, Michael Chui, Brad Brown, Jacques Bughin, Richard Dobbs, Charles Roxburgh, and Angela H Byers. Big data: The next frontier for innovation, competition, and productivity. 2011.

MMSB15.   Mahshad M Mahally, Miroslaw Staron, and Jan Bosch. Barriers and enablers for shortening software development lead-time in mechatronics organizations: A case study. In *Proceedings of the 2015 10th Joint Meeting on Foundations of Software Engineering*, pages 1006–1009. ACM, 2015.

MNSKS05. Manoel, E., Nielson, M.J., Salahshour, A., KVL, S.S. and Sudarshanan, S., 2005. *Problem determination using self-managing autonomic technology*. IBM International Technical Support Organization.

MS04.   Peter Manhart and Kurt Schneider. Breaking the ice for agile development of embedded software: An industry experience report. In *Proceedings of the 26th international Conference on Software Engineering*, pages 378–386. IEEE Computer Society, 2004.

MSC13.   Viktor Mayer-Schönberger and Kenneth Cukier. *Big data: A revolution that will transform how we live, work, and think.* Houghton Mifflin Harcourt, 2013.

Pas14.   A Pasztor. Tesla unveils all-wheel-drive, autopilot for electric cars. *The Wall Street Journal*, 2014.

PW92.   Dewayne E Perry and Alexander L Wolf. Foundations for the study of software architecture. *ACM SIGSOFT Software Engineering Notes*, 17(4):40–52, 1992.

Ric11.   Eric Richardson. What an agile architect can learn from a hurricane meteorologist. *IEEE software*, 28(6):9–12, 2011.

Rie11.   Eric Ries. *The lean startup: How today's entrepreneurs use continuous innovation to create radically successful businesses.* Random House LLC, 2011.

SBB+09.   Helen Sharp, Nathan Baddoo, Sarah Beecham, Tracy Hall, and Hugh Robinson. Models of motivation in software engineering. *Information and Software Technology*, 51(1):219–233, 2009.

Sin11.   Purnendu Sinha. Architectural design and reliability analysis of a fail-operational brake-by-wire system from iso 26262 perspectives. *Reliability Engineering & System Safety*, 96(10):1349–1359, 2011.

SLS+13.   Florian Sagstetter, Martin Lukasiewycz, Sebastian Steinhorst, Marko Wolf, Alexandre Bouard, William R Harris, Somesh Jha, Thomas Peyrin, Axel Poschmann, and Samarjit Chakraborty. Security challenges in automotive hardware/software architecture design. In *Proceedings of the Conference on Design, Automation and Test in Europe*, pages 458–463. EDA Consortium, 2013.

SM11.   Miroslaw Staron and Wilhelm Meding. Monitoring Bottlenecks in Agile and Lean Software Development Projects–A Method and Its Industrial Use. *Product-Focused Software Process Improvement*, pages 3–16, 2011.

SRH15.   Miroslaw Staron, Rakesh Rana, and Jörgen Hansson. Influence of software complexity on iso/iec 26262 software verification requirements. 2015.

SS16.   Miroslaw Staron and Riccardo Scandariato. Data veracity in intelligent transportation systems: the slippery road warning scenario. In *Intelligent Vehicles Symposium*, 2016.

Wri11.   Alex Wright. Hacking cars. *Communications of the ACM*, 54(11):18–19, 2011.

Chapter 10
Summary

Abstract In this book we have introduced the concept of *software architecture* in automotive software and overviewed different architectural styles that can be encountered in modern automotive software. In this chapter we present the summary of the main points of the book and pinpoint additional reading in the area.

10.1 Software Architectures in General and in the Automotive Software: A Short Recap

Software architecture is a high level design and organization of the software system. It provides guidelines for the detailed design of the software, its components and their deployment. It is usual that software architecture documentation contains a number of different viewpoints, such as the functional viewpoint, the logical viewpoints, or the deployment one.

As the software architecture also provides the principles of the high-level organization of the software system, they often include different architectural styles. In general, we could observe over 20 different styles which are often accompanies by over 20 different patterns. However, in the automotive software design only some of these styles and patterns are applicable.

In this book we collected the most important methods and tools for the design of automotive software—both at the architectural level and at a detailed design level. In this chapter we provide a short summary of each chapter and briefly outline why this knowledge is important for the future of the automotive software engineering.

10.2 Chapter 2: Software Architectures

In the second chapter of this book we introduced the notion of software architecture as *high level structures of a software system, the discipline of creating such structures, and the documentation of these structures.* We have introduced the notion of software component and discussed a set of architectural viewpoints which are

© Springer International Publishing AG 2017

M. Staron, *Automotive Software Architectures*,

DOI 10.1007/978-3-319-58610-6_10

common in the design of the automotive software systems, such as:

- Functional view—describing the architecture of the functions of the vehicle and the dependency between them
- Physical view—describing the physical nodes (ECUs) and their connections
- Logical view—describing the software components and their organization, and
- Deployment view—describing the deployment of software components onto the ECUs

We have also exemplified the main architectural styles present in the automotive sector:

- Layered architecture
- Component-based architecture
- Monolithic architecture
- Microkernel architecture
- Pipes and filters architecture
- Event-driven architecture, and
- Middleware architecture with message brokers

The knowledge contained in Chap. 2 prepares us to start designing software systems at a very high level. In order to be effective we need to understand how the automotive software development is done, and therefore we describe it in the next chapter.

10.3   Chapter 3: Automotive Software Engineering

When describing the automotive software engineering practices, we start with the description of requirements, which are a bit specific for the automotive software. We discuss the following types of requirements:

- Textual requirements—specifications which are done in form of free text or tables
- Use case requirements—specifications which are based on UML Use cases and the corresponding sequence diagrams
- Model-based requirements—specifications which are done in form of models that should be implemented by suppliers

We need to understand the way in which requirements are done so that we understand the way in which software verification and validation is done. This verification and validation, done in form of testing, is discussed in the remaining of that chapter, where we introduce:

- Unit testing—verification of the functionality of individual software modules
- Component testing—verification of groups of software modules—components
- System testing—verification of the complete system (both of its complete functionality and partial functionality during the course of the development), and

- Functional testing—validation of the end user functions against their specifications

Once we introduce the different testing techniques, and the stages of integration of the software of a car, we discuss how these elements are stored in the so-called product databases.

10.4   Chapter 4: AUTOSAR

One of the major trends in today's automotive software is the introduction of the AUTOSAR standard. The standard specifies how the automotive software is to be organized and how it should communicate with each other. By some, the standard is viewed as "the operating system" of the car.

This chapter has been written by Darko Durisic, who is one of the representatives of one of the Swedish OEMs in the AUTOSAR consortium. His research and expertise in the area resulted in a good introduction to the standard from the perspective of a software designer. The focus on the chapter is on the reference architecture provided by AUTOSAR and its implications.

In this chapter we also discuss the evolution of the AUTOSAR from the perspective of software concepts—which elements of the specification have changed, how the number of concept has evolved and what it means for the automotive software design.

We conclude the chapter by providing a few examples of how to design the modern software components based on AUTOSAR.

10.5   Chapter 5: Detailed Design of Automotive Software

The discussion about automotive architectures would not be complete if we did not discuss the methods for detailed design of this software. In Chap. 5 we introduce a number of methods:

- Simulink modelling—probably the most widely used method for detailed design of algorithms in the automotive software, usually used in such domains as Powertrain and Active Safety or Chassi development.
- SysML—a UML based method for specifying the software focused on the concepts of the programming languages.
- EAST-ADL—another UML based method for designing the automotive software, combining the problem domain concepts with the programming/system level concepts.
- GENIVI—a standard for programming infotainment systems, which is currently gaining increasingly more popularity in the market.

Knowing the notation is one thing, understanding the principles of the design of safety-critical systems is another. Therefore we introduce the principles of designing of safety critical systems based on the research from NASA and its space program.

10.6 Chapter 6: Evaluation of Automotive Software Architectures

Once we introduce the detailed design we also discuss methods for evaluating software architectures. In Chap. 6 we focus on presenting methods based on qualitative evaluations—we focus on Architecture Trade-Off Analysis Method (ATAM).

We start the chapter by introducing the rationale for the evaluation of the architectures anchored in the international standard ISO/IEC 25000. We then describe the ATAM and provide a number of typical scenarios for evaluating safety critical systems.

Finally we present an example evaluation of a simple architectural design.

10.7 Chapter 7: Metrics for Software Designs and Architectures

To complement the methods presented in Chap. 6, we focus on methods based on quantitative measurements of software design. We introduce the international standard ISO/IEC 15939 for software measurement processes and we discuss abstraction levels of different metrics.

We provide a set of measures used by software architects—an architect portfolio—and their visualizations. We also present a set of metrics for the detailed designs of the automotive software.

As an example in this chapter we present measurement results of publicly available industrial data set from one of the modern cars. Based on this open data we discuss the properties of the software such as its size or cyclomatic complexity. We reason what that means for the validation of software and its safety.

This chapter is co-authored with Wilhelm Meding from Ericsson, who is a senior measurement program and team leader and has been working in this domain for more than 10 year.

10.8 Chapter 8: Functional Safety of Automotive Software

Once we outlined the risks of not being able to fully validate the software once it becomes too complex, we move on and introduce one of the major standard in the automotive software today—ISO/IEC 26262 (functional safety).

This chapter was authored by Per Johannessen from Volvo AB, who was part of the introduction of the ISO/IEC 26262 standard to one of the OEMs of the passenger vehicles and is currently working on the same topic for heavy vehicles and buses.

As an example in this chapter we present an architecture of a microcontroller where the different ASIL levels are demonstrated.

10.9 Chapter 9: Current Trends

Finally we close this book by outlining the current trends in the automotive software development. We outline the following trends:

- Autonomous driving—a trend which requires more complex software and higher degree of connectivity
- Self-healing, self-adaptive, self-organizing systems—a trend which enables more reliable and smarter software, but is challenging in terms of safety assessment over time
- Big data—a trend which enables the cars' software to make smarter decisions based on the availability of information from external sources, at the same time putting requirements on the processing power, storage and other characteristics of the software system
- New trends in software development—for example the trend of continuous integration of software which enables constant improvements of the software, but at the same time putting a lot of requirements on safety assessment and validation of the software on-the-fly

10.10 Closing Remarks

At this point we finish our journey through the automotive software development of the second decade of the second millennium. We see that we live in very dynamic times where the field of software engineering in the automotive sector just starts to grow and expand rapidly.

We hope that this book will help You, the reader to become a better software engineer and will help the cars to be smarter, better, more fun and above all—safer!

CPSIA information can be obtained
at www.ICGtesting.com
Printed in the USA
LVHW021818210419
614974LV00002B/12/P